KINGS FOR BID WINNING
—CREATIVE CONCEPTION
得标为王——创意篇
2014

龙志伟 编著
Edited by Long Zhiwei

城市综合体	商业建筑	酒店建筑	交通建筑	文化艺术建筑
学校建筑	办公建筑	住宅建筑	综合
Urban Complex	Commercial Building	Hotel	Transportation Building
Culture & Art Building	School Building	Office Building	Residence	Others

广西师范大学出版社
·桂林·

图书在版编目(CIP)数据

得标为王:创意篇2014/龙志伟 编著. —桂林:广西师范大学出版社,2014.6
ISBN 978-7-5495-5164-4

Ⅰ. ①得… Ⅱ. ①龙… Ⅲ. ①建筑设计-作品集-世界-现代 Ⅳ. ①TU206

中国版本图书馆CIP数据核字(2014)第042625号

出 品 人:刘广汉
责任编辑:王晨晖
装帧设计:龙志杰

广西师范大学出版社出版发行

(广西桂林市中华路22号　邮政编码:541001)
(网址:http://www.bbtpress.com)

出版人:何林夏
全国新华书店经销
销售热线:021-31260822-882/883
上海锦良印刷厂印刷
(上海市普陀区真南路2548号6号楼　邮政编码:200331)
开本:646mm×960mm　1/8
印张:47　　　　　字数:25千字
2014年6月第1版　　2014年6月第1次印刷
定价:348.00元

如发现印装质量问题,影响阅读,请与印刷单位联系调换。
(电话:021-56519605)

■ 序 Preface

如若说，在创新尚属于人类个体或群体中的个别杰出表现时，人们循规蹈矩的生存姿态尚可为时代所容，那么，在创新将成为人类赖以进行生存竞争的不可或缺的素质时，依然采取一种循规蹈矩的生存姿态，则无异于一种自我溃败。毋庸置疑，当今我们所处的时代，正是一个鼓励创新、崇尚创新的时代，创新已然成为推动个人、企业、各行业乃至国家和整个社会的关键。

《得标为王——创意篇2014》就是这么一本讲述设计师如何通过多学科、多专业结合以及跳跃性的创造性思维衍生出惊人创意的建筑专著。本书从思维的高度和广度统领全书，从城市综合体、商业建筑、酒店建筑、交通建筑、文化艺术建筑、学校建筑、办公建筑、住宅建筑、综合等类别，重点讲述了设计师创作的创意来源、构思思路以及相应的解决途径，并辅以表达设计师创意来源的手绘图和精细的技术图，让读者能够深入到设计师的思维中，与之产生思想上的共鸣！

翻阅本书，你将仰望并追逐阿特金斯、DnA_Design and Architecture、斯蒂文·霍尔、Henning Larsen Architects、C.F. Møller、HKG、LAVA、绿舍都会等名家身影。细品此书，你将置身于一个惊人的创意世界：在这里，体验视觉的震撼，感受大自然和生活中简单、平凡的事物折射的光辉；在这里，接受思维的洗礼，跟随设计师的思绪，见证创意的诞生过程，提升思维的高度。

A formalist life can only survive and be accepted in an era when innovation is exclusive to certain individuals or groups; living a rigid life when innovation becomes an indispensable quality is nothing but self defeat. Our time encourages and advocates innovation. Innovation has become crucial to push individuals, enterprises, industries even countries forward.

Kings for Bid Winning – Creative Conception 2014 tells about how designers create innovative buildings with their interdisciplinary knowledge and creative thinking. From the categories of Urban Complex, Commercial Building, Hotel, Transportation Building, Culture & Art Building, School Building, Office Building, Residence and Others, the book emphatically represents the source of innovative ideas, conceptions and corresponding solutions. Hand drawings and delicate technical drawings enable readers to get into designers' mind and provoke a resonance of ideas.

Projects from Atkins, DnA_Design and Architecture, Steven Holl Architects, Henning Larsen Architects, C.F. Møller, HKG, LAVA, SURE Architecture etc. are included. You are about to enter a creative world: in here, you will experience visual impact, enjoy brilliance from the nature and simple ordinary things of life, meet new ideas, and witness the birth of innovative ideas by following the designers.

目录 Contents

8 城市综合体
Urban Complex

10 广东深圳雅宝高科技企业总部园
Yabao Hi-Tech Enterprises Headquarters Park

18 江苏常州揽月湾西地会议中心及综合发展项目
Lanyue Bay West Convention Center & Mixed Development

24 阿联酋阿布扎比Corniche塔楼
Corniche Tower

34 广东珠海华融横琴大厦
Zhuhai Huarong Hengqin Tower

40 商业建筑
Commercial Building

42 广东珠海世邦国际商贸中心
Summer International Retail and Entertainment Center

48 土耳其布尔萨蔬果和鱼类批发市场
Bursa Wholesale Greengrocers and Fishmongers Market

56 墨西哥合众国墨西哥城佩德雷加尔购物中心
Pedregal Shopping Center

62 伊朗德黑兰证券交易所
Teheran Stock Exchange

68 荷兰阿尔梅勒世贸中心
WTC Center Almere

72 酒店建筑
Hotel

74 拉脱维亚里加新Liesma酒店
Hotel Liesma

82 黑山共和国Rafailovichi Apart酒店及地下停车场
Rafailovichi Apart-hotel & Underground Parking

88 阿联酋阿布扎比埃米尔珍珠酒店
Regent Emirates Pearl

94 格鲁吉亚巴统Medea酒店
Medea Hotel

100 浙江杭州西溪度假酒店
Hangzhou Xixi Tourism Resort Hotel

110 交通建筑
Transportation Building

112 哈萨克斯坦阿斯塔纳火车站
Astana Railway Station

116 意大利那不勒斯Garibaldi广场
Piazza Garibaldi

120 台湾高雄港和游轮中心
Kaohsiung Port and Cruise Center

126 黎巴嫩贝鲁特码头 & 城镇港口
Beirut Marina & Town Quay

136 文化艺术建筑
Culture & Art Building

138 荷兰埃曼剧院及动物园
Emmen Theater and Zoo

144 埃塞俄比亚亚的斯亚贝巴国家体育馆和体育村
National Stadium and Sports Village

150 台湾新北艺术博物馆
New Taipei City Museum of Art

154 美国纽约韩国文化中心
Korean Cultural Center in New York

164 韩国釜山歌剧院
Busan Opera House

172 美国犹他州帕克城金博尔艺术中心
Kimball Art Center

178 韩国大邱高山郡公共图书馆
Daego Gosan Public Library

186 俄罗斯南萨哈林斯克商业和展览中心
Business & Exhibition Center in Yuzhno-Sakhalinsk

192 重庆"城市森林"
Chongqing "Urban Forest"

196 学校建筑
School Building

198 丹麦奥尔胡斯纳维达斯公园工程学院和科学公园
Navitas Park Engineering College & Science Park

210 瑞士洛桑市洛桑联邦理工学院教学桥整修及机械学大厅和图书馆扩建
Teaching Bridge at the Ecole Polytechnique Fédérale de Lausanne /
Rehabilitation – Extension of the Mechanics Hall and Library

216 福建厦门光电职业技术学院
The Fujian Professional Photonic Technical College

224 办公建筑
Office Building

226 格鲁吉亚巴统法院
Palace of Justice of Batumi

234 日本东京比利时驻日大使馆改建
Belgian Embassy Reconstruction

244 葡萄牙Maia LÚCIO新总部办公大楼
LÚCIO's New Headquarters Office Building

254 伊朗布什尔建筑工程组织大楼
Bushehr Construction Engineering Organization Building

260 西班牙安达卢西亚自治区政府办公楼
Office Tower for the Government of Andalusia

268 住宅建筑 Residence

- **270** 瑞典哥特堡Harbor Stones住宅区 Harbor Stones
- **274** 奥地利下奥地利州樱桃院住宅 Cherry Yard's House
- **280** 瑞典斯德哥尔摩Alvik大楼 Alvik Tower
- **286** 荷兰皮尔默伦德Londenhaven公寓大楼 Londenhaven
- **294** 秘鲁利马精品公寓 Loft Boutique
- **302** 墨西哥合众国瓦拉塔港La Peninsula La Peninsula

306 综合 Others

- **308** 台湾台中"台湾塔" Taiwan Tower
- **314** 保加利亚索菲亚Collider活动中心 Collider Activity Center
- **318** 希腊雅典斯帕塔阿提卡公园Okeanopolis水族馆 Athens Spata Attica Park Okeanopolis Aquarium
- **326** 浙江杭州西溪休闲中心 Xixi Leisure Center
- **330** 荷兰乌特勒支Belle van Zuylen大楼 Belle van Zuylen Tower
- **342** 丹麦Køge大学医院 Køge University Hospital
- **358** 墨西哥合众国墨西哥城"垂直公园" Vertical Park
- **362** 墨西哥合众国库利亚坎Nativity教堂 Nativity Church

城市综合体 Urban Complex

绿色湖岛
"溪水涌泉"
立体花园

Green Lake Island, "Stream Fountain", Three-dimensional Garden

广东深圳雅宝高科技企业总部园
Yabao Hi-Tech Enterprises Headquarters Park

设计单位：10 DESIGN（拾稼设计）
开发商：星河集团
项目地址：中国广东省深圳市
占地面积：650 000 ㎡
建筑面积：1 050 000 ㎡
建筑团队：Ted Givens　Maciej Setniewski
　　　　　Peby Pratama　Tatsu Hayashi
　　　　　Abraham Fung　Emre Icdem
　　　　　Ru Chen
景观团队：Ewa Koter　Ting Fung Chan
多媒体制作：Shane Dale　Jon Martin

Designed by: 10 DESIGN
Client: Galaxy Group
Location: Shenzhen, Guangdong, China
Site Area: 650,000 m²
Gross Floor Area: 1,050,000 m²
Architecture Team: Ted Givens, Maciej Setniewski, Peby Pratama, Tatsu Hayashi, Abraham Fung, Emre Icdem, Ru Chen
Landscape Team: Ewa Koter, Ting Fung Chan
CGI: Shane Dale, Jon Martin

项目概况

雅宝高科技企业总部园区靠近深圳市福田区的中心地带，占地65公顷，总建筑面积为1 050 000 ㎡，由18座高层办公楼、1座五星级酒店、3座酒店式公寓、3栋住宅楼、1座购物中心以及32公顷的停车场组成。

设计构思

项目是对原始乡村景观与城市极速发展动力之间的关系的探讨。基地处于茂密树丛间的开阔草地上，四周由两个天然湖泊环绕，考虑到这一独特的地貌特征，设计以将项目与当地的自然风光融合为主要设计理念，同时，结合周边一流的配套设施，旨在构建一座宁静而又富有创意的高科技企业总部园区，平衡工作与生活之间的关系。

设计特色

建筑群在基地西南面的边缘处界定了城市界限，并沿着东北方向穿越基地，延展至湖泊处，逐渐融合在自然景观中。塔楼上的一组组阳台拉近了建筑外墙与大自然的距离，使植物能够沿着建筑墙面生长；一系列攀附于建筑西立面的海藻管也赋予了建筑群绿意；建筑外部线形屏幕的设计进一步模糊了塔楼的轮廓，并保护塔楼免受夏季烈日的暴晒，而每栋建筑顶部的屋顶花园也有效地缓解了城市热岛效应。

项目在建筑造型上极富创意，形成了两座地标式建筑。第一座高300 m，位于穿越基地的溪流边上。受溪水流动性的启发，这栋建筑楼体呈螺旋形上升，好像涌出地面的溪水。第二座标志性建筑为这个区域的购物中心，高220 m，坐落于两条高速公路的交叉口，与高速公路相呼应，横向拉近了高速路与建筑群的距离，将塔楼、购物中心融入了400 m范围的标志性建筑群。购物中心的内向外缘由阶梯状露台组成，一直通往绿色花园空间。

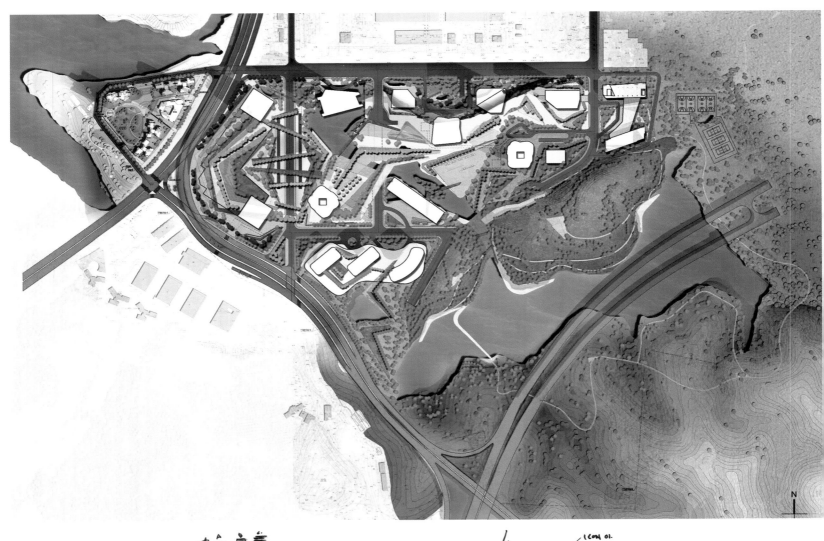

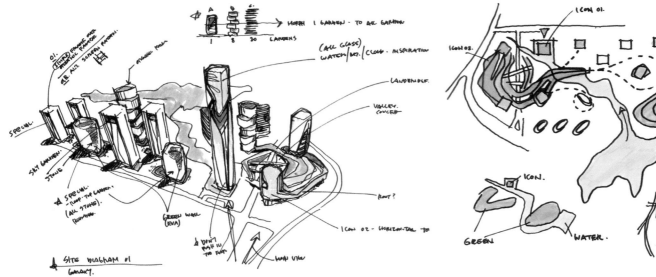

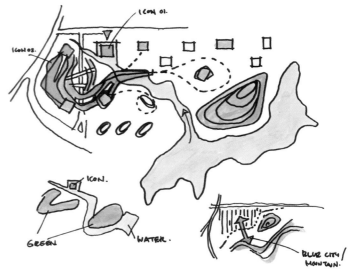

Landscape Concept Strategies
景观概念策略

STEP 1. TAKING THE ENERGY FROM THE EXISTING LANDSCAPE MOUNTAINS
步骤一：从已有景观山体中获得能量

STEP 2. ADDING LANDSCAPE LAYERS WHICH CREATING CONNECTING SMALLER SPACES TO ALLOW PROGRAM FOR DIVERSE USERS
步骤二：新增景观层，打造小型连接空间以迎合多样化的用户需求

STEP 3. CONFLUENCE LANDSCAPE AND ARCHITECTURE BY REFLECTING LANDSCAPE AND CREATING A NEW GREEN TEXTURE
步骤三：通过反映景观和打造一个新的绿色肌理将景观和建筑结合起来

STEP 4. CONNECTING THE WATER ENERGY THROUGH SITE AND COMPLIMENT SELECTED ARCHITECTURE
步骤四：将场地内的水能与建筑结合

Profile

Yabao Hi-Tech Enterprises Headquarters Park is close to the central zone of Futian District, Shenzhen, China. The gross site area is about 65 ha; and the GFA is over 1,050,000 square meters, consisting of 18 high-rise towers, a 5 star hotel, 3 service apartment towers, 3 residential towers, a shopping mall and a 32 ha parking lot.

Design Concept

The project is an examination of the relationship between a pristine rural landscape and the advancing forces of a rapidly growing city. The site is very unique with superb first world infrastructure in beautiful parkland setting around two natural lakes. The main design concept was to try and integrate the complex into the natural landscape. The vision for the project is to cultivate a hi-tech headquarter park where people can lead a well-balanced life, working and living in a tranquil yet innovative environment.

Design Feature

The buildings define a strong urban edge on the southwestern edge of the site and begin to dissolve into nature as they move northeast across the site towards the lake. A series of balconies pulls off the tower façades to allow for vegetation to grow up the sides of the buildings. A series of algae tubes mounted on the western façades also brings a natural diffuse green texture into the complex. External linear screens further diffuse the edge of the towers and shelter the buildings from the heat of the summer sun. Each tower has a rooftop garden to help reduce heat island effect.

The architecture characterized by creative shape is defined by two landmark towers. The first is a 300 meter tower that sits at the edge of a small stream running across the site. The tower twists up out of the stream taking inspiration from the fluidity of the water running across the site. A second iconic element is the shopping mall, which is located at the intersection of two freeways. To react to the freeway edge, a 220 meter tower is pulled laterally along the freeway, melding the tower and shopping mall into an iconic element that reaches 400 meters in length. The internal edge of the shopping mall composed of terraces leads to a series of green garden spaces.

Main Tower
主塔楼

Retail Mall
零售商场

Special Office 1
特别办公室 1

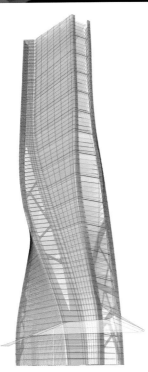

Main Tower
主塔楼

Retail Mall
零售商场

Special Office
特别办公室 1

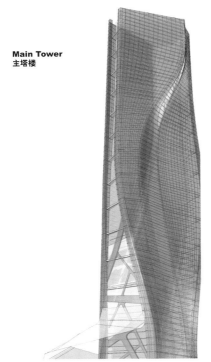
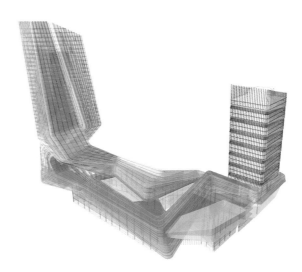

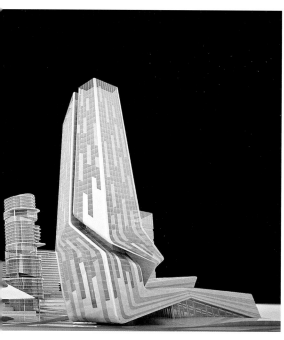
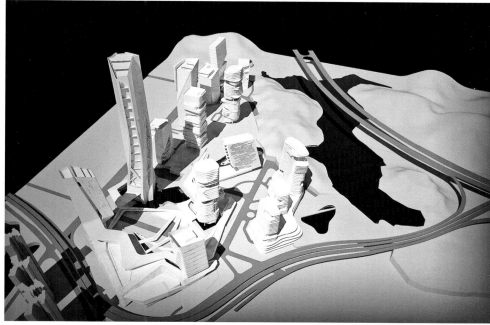

Special Office 2
特别办公室 2

Special Office 3
特别办公室 3

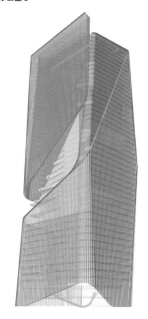
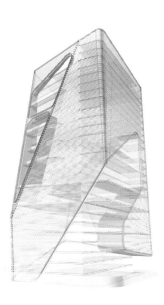

Special Office 2
特别办公室 2

Special Office 3
特别办公室 3

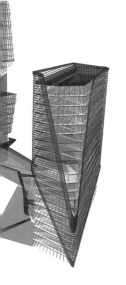
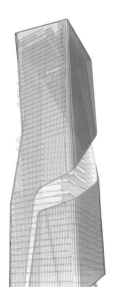
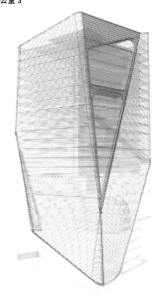

江苏常州揽月湾西地会议中心及综合发展项目
Lanyue Bay West Convention Center & Mixed Development

设计单位：10 DESIGN（拾稼设计）	Designed by: 10 DESIGN
开发商：星河集团	Client: Galaxy Group
项目地址：中国江苏省常州市	Location: Changzhou, Jiangsu, China
占地面积：132 298 ㎡	Site Area: 132,298 m²
总建筑面积：304 285 ㎡	Gross Floor Area: 304,285 m²
设计团队：Ted Givens Adam Wang Adrian Yau Audrey Ma Peby Pratama Yao Ma	Design Team: Ted Givens, Adam Wang, Adrian Yau, Audrey Ma, Peby Pratama, Yao Ma

项目概况

项目位于常州市武进区滨湖新城的西太湖生态休闲区，将成为长江三角洲地区一个重要的旅游度假区。包括总建筑面积在内的项目总面积达 31 000 000 m²，将规划形成一个现代化的、可持续性的休闲和商业社区，其发展业态包括一个标志性的豪华酒店、酒店式公寓、会议中心、办公楼及公共空间等。

设计构思

项目位于揽月湾 14 km长的海岸线上，项目基地的环境激发了设计师通过花园环境来提升滨湖公共空间品质的设想。设计师从当地的竹雕刻工艺及具有流线型的现代船舶设计中获取灵感构建这个项目。建筑物呈现出流畅的弧形结构，既柔化了建筑线条，又与揽月湾水体的流动性相呼应。每座塔楼的剖面均以雕刻手法塑造大型的海上漂浮花园，使整体环境更加柔和舒缓，给人置身于度假村之感。

绿色设计

建筑设计充分考虑了基地特征与周边环境，使三者互相呼应，相得益彰。设计认真考虑了周边建筑物的布局，使项目在最大限度内实现了自然通风。绿地空间的进一步引入，既彰显了绿色设计的概念，同时也营造了适宜的微气候。可持续设计还体现在酒店塔楼外墙的设计上，该外墙采用了二氧化钛纳米涂料，这种涂料可以中和空气中的污染物，清洁空气，这将使该建筑成为常州清洁空气的象征性建筑。

Bamboo Carving,
Modern Boat,
Floating Garden

竹雕工艺
现代船舶
海上漂浮花园

Profile

The project is located in the West Tai Lake's Ecological and Leisure Area, positioned to be one of the most important vacation resorts in the Delta Region of Yangtze River. The 31,000,000 square meters of development will form a modern leisure and commercial community, providing an iconic luxury hotel, serviced apartments, conference centers, offices and public spaces.

Design Concept

Given the unique geography of the site, situating along the 14 km coastline of Lanyue Bay, designers envisaged to create a garden setting environment in order to help increasing the public spaces surrounding the lake. The masterplan design was inspired by the local craft of bamboo carving and the aerodynamic shapes of modern boat design. The buildings are all curved to reflect the fluidity of the lake. A section of each tower is carved to house a large floating garden to give the masterplan a softer, more resort like feel.

Green Design

Architectural design and planning arrangement respond both to its surrounding environment and the site itself. Careful consideration was paid to zoning and building arrangement within the site to maximize views and ventilation to all buildings. The introduction of green spaces further implements the concept of green design and creates enjoyable micro-climate. As an additional sustainable concept the hotel tower can be clad in a titanium dioxide nano-coating. The coating neutralizes air pollution and cleans the air. The tower can become an iconic symbol of the clean air of Changzhou.

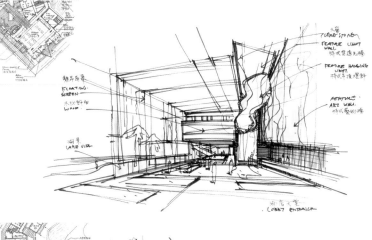
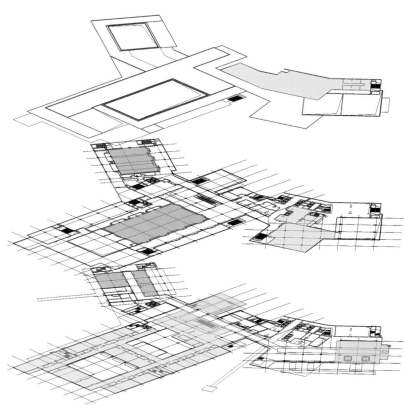
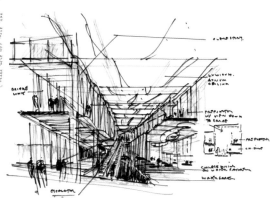

"万花筒"
招风斗
流动光束

"Kaleidoscope",
Wind Catcher,
Light Beam

阿联酋阿布扎比 Corniche 塔楼
Corniche Tower

设计单位：Oppenheim Architecture+Design
项目地址：阿联酋阿布扎比
项目面积：157 935 ㎡

Designed by: Oppenheim Architecture + Design
Location: Abu Dhabi, UAE
Area: 157,935 m²

设计构思

为了呼应场地的几何形态，设计师将这个坐落在阿联酋阿布扎比一个显著场地上的多功能塔楼设想为一个有着通透外观的圆柱形建筑体量，层层叠叠的圆形空间聚集起来，逐渐漂浮在"透明薄膜"内，宛如公园内被捕捉到的光束，又像一个缤纷万花筒一般，模糊了公园、天空和水体的界限。

设计特色

整个项目主要分为 4 个功能区：零售、办公、酒店和住宅。为了满足开放性的设计需求，这 4 个功能区所占的比重可根据需求而进行灵活的调整，其中任一"圆盘"的增加或删减甚至去除，都不会影响建筑最初设定的主体形态。同样，若政府规定不允许使用建筑红线外的地下区域，停车场和零售区也可以进行相应的重新配置。

建筑体环路的堆叠围绕着一个可伸缩的核心系统来组织，确保建筑内拥有可通往每一个功能区的独立且便捷的通道，而环路与环路之间的过渡空间则由绿色休闲环路支撑。可伸缩核心部分使公园可依据功能需求扩展为大型的中庭，成为办公区上方连续的公园区和采光区。

可持续性设计

设计师非常注重建筑的可持续性和对自然资源的利用。建筑的形态参照了当地传统的招风斗，建筑本身就是一个可将凉爽的微风引导至室内的空间装置；建筑的顶端设置有可利用风能和太阳能的装置，通过对自然能源的利用，缓解建筑的能源需求；对流经场地的水体的利用，也有利于建筑实现能源的自给。

SITE PLAN
总平面图

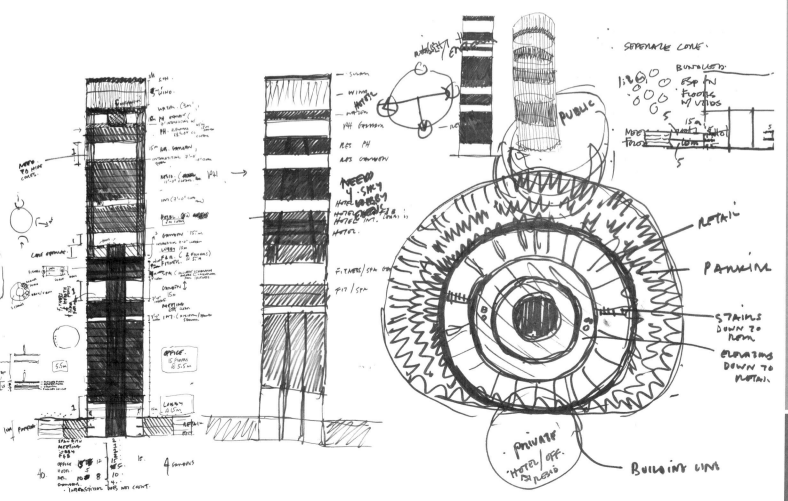

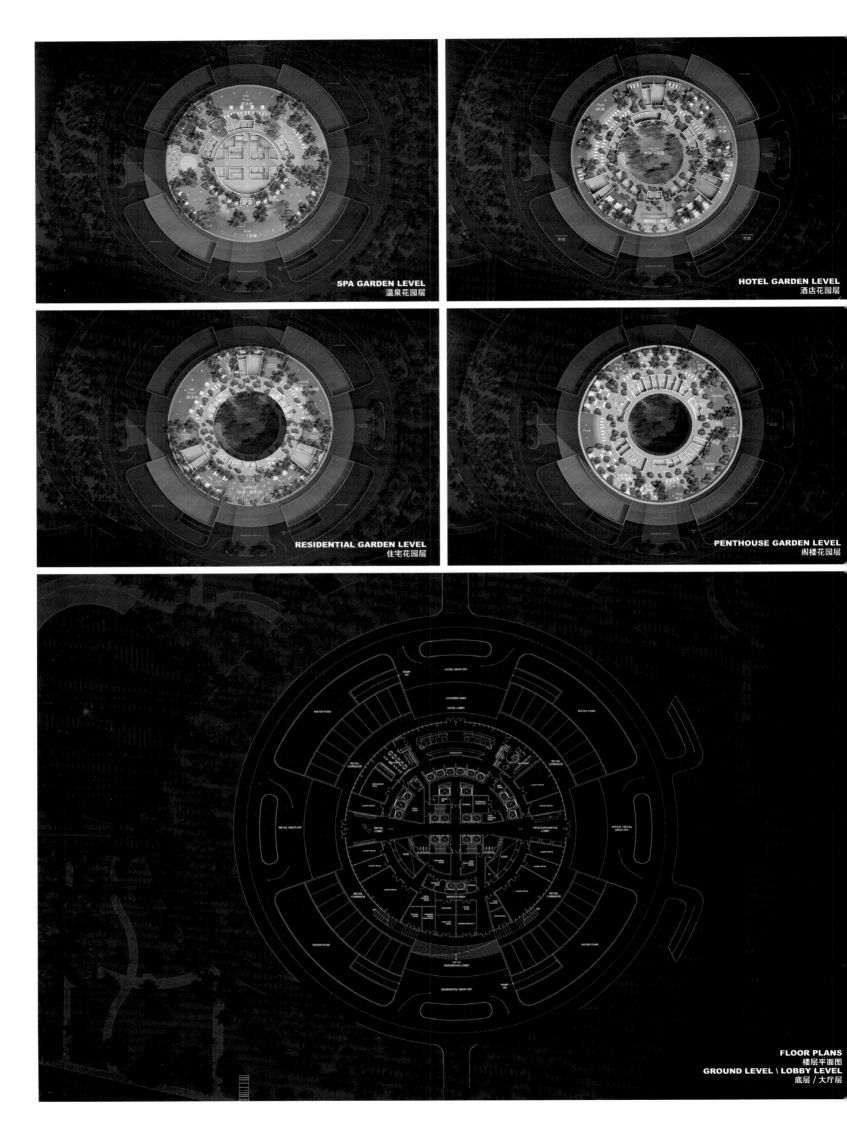

MEZZANINE LEVEL
夹层

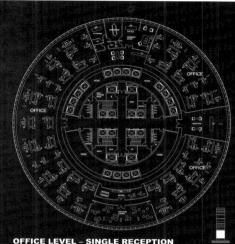
OFFICE LEVEL – SINGLE RECEPTION
办公层——单接待处

OFFICE LEVEL – DOUBLE RECEPTION
办公层——双接待处

MEETING AREAS LEVEL
会议区水平面

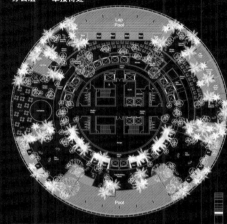
SPA GARDEN LEVEL
温泉花园层

SPA LEVEL
水疗层

FLOOR PLANS
楼层平面图

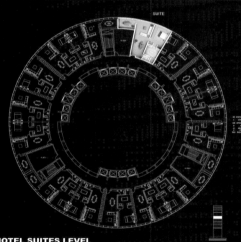
HOTEL SUITES LEVEL
酒店套房层

RESIDENTIAL LEVEL – 9 UNITS
住宅层——9个单元

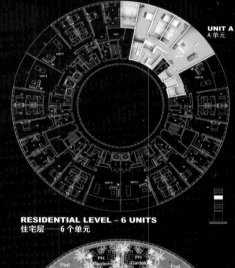
RESIDENTIAL LEVEL – 6 UNITS
住宅层——6个单元

RESIDENTIAL GARDEN LEVEL
住宅花园层

RESIDENTIAL PENTHOUSE LEVEL
住宅阁楼层

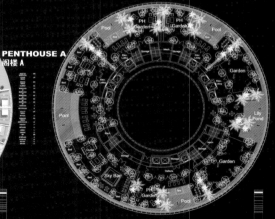
PENTHOUSE GARDEN LEVEL
阁楼花园层

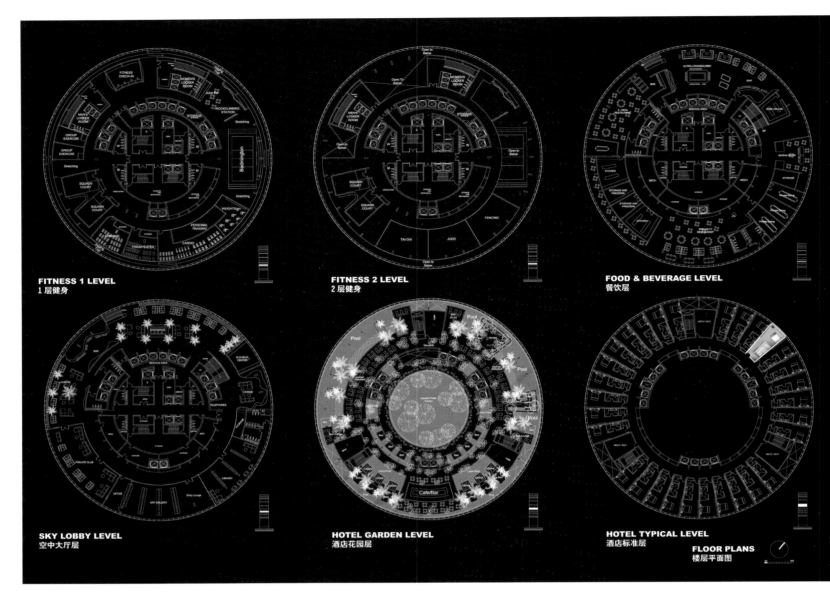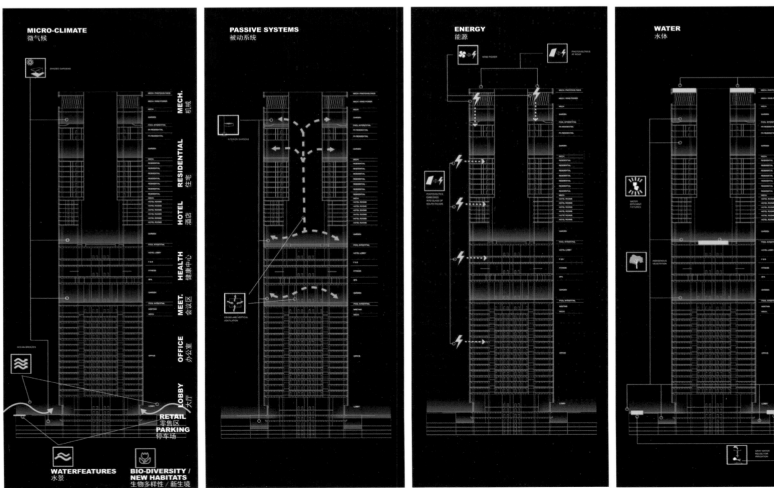

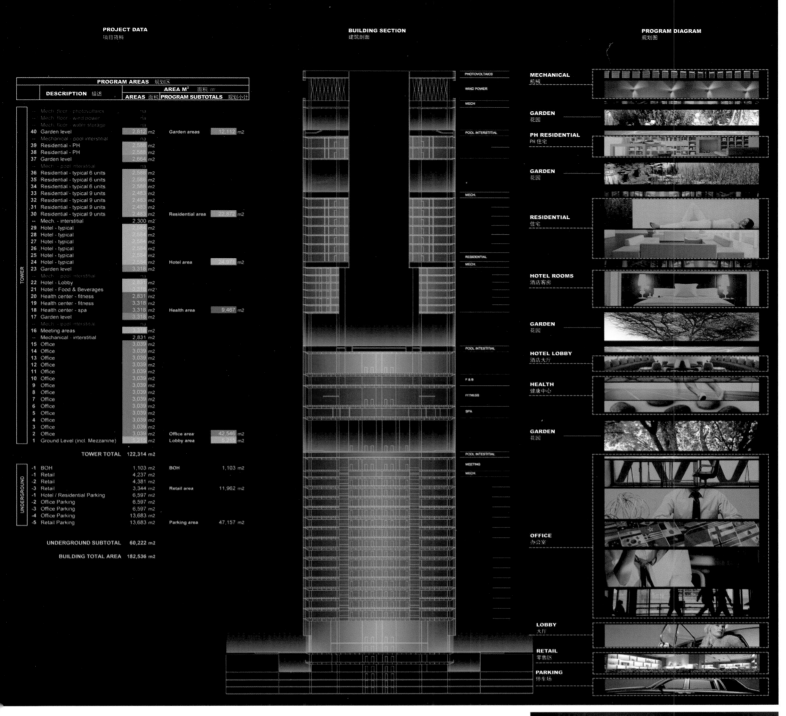

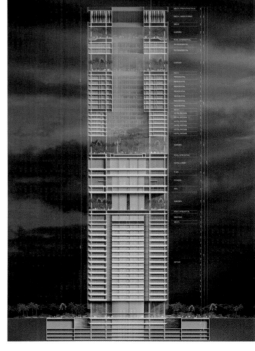

Design Concept

As a response to the geometry of the site, designers envisage this multi-functional tower in Abu Dhabi, UAE as a cylindrical building volume with transparent appearance. The building emerges elementally, as a stacked assemblage of program floating within a diaphanous membrane, just like a captured beam of light solidified in the park blurring distinctions between park, sky and water.

Design Feature

The project is divided into four functional quadrants: retail, office, hotel, and residential programs. Pertinent to the open-ended nature of the programming requirements, the proportion of the four parts can be flexibly adjusted according to specific demands, without affecting in the least the form of the initial proposition. Similarly if the municipal dictates disallow use of subterranean areas beyond the property lines the parking and retail program could be reconfigured accordingly.

Each ring of program within the stack is organized around a system of telescoping cores that allows direct and independent access to each of the functions. Each ascending ring of program has been buoyed at each transition by verdant leisure rings. As a result of the telescoping cores, the gardens are allowed to expand into a vast atrium, creating a continuous column of gardens and light, above the office functions.

Sustainable Design

An integral aspect of this orchestration has been the focus on self sustainability and an embrace of the natural resources. The building form itself becomes a device for luring cool breezes to spaces shielded from the sun, referencing the regional time-tested tradition of wind catchers. Inconspicuously and opportunistically the structure incorporates solar and wind arrays for the generation of some of this building's required energy. Other integral systems deployed towards complete self sustenance, include various methods for the reuse of the vast amounts of water flowing through the site.

广东珠海华融横琴大厦
Zhuhai Huarong Hengqin Tower

设计单位：阿特金斯 ATKINS	**Designed by: Atkins**
开发商：华融置业有限公司	**Developer: Huarong Real Estate Co., Ltd.**
项目地址：中国广东省珠海市	**Location: Zhuhai, Guangdong, China**

流线体量
波浪百叶
太极八卦图

Linear Volume,
Wavy Louver,
Eight Trigrams

项目概况

华融横琴大厦是一个综合用途发展项目，包括五星级酒店、国际甲级办公楼及高级商业配套等功能，将成为横琴新经济特区总体规划的首批建筑物之一。

设计构思

依据基地现有环境，设计师旨在提供一个醒目、现代且符合开发商愿景的设计方案。为使酒店客人及办公楼用户均能眺望澳门地标性建筑，设计师对大厦进行了精心的设计。设计师从太极八卦图获取灵感，设想了两个线条流畅、相对而立的建筑体量，两栋建筑之间因错位而形成开敞空间，保证建筑可享有开阔的视野。设计也充分考虑了建筑与环境的关系，这一独特的建筑造型为城市的未来开发预留了空间，不会对已有或潜在发展项目造成视觉干扰。

考虑到沿海地区的气候，设计师将建筑设想为特殊的流线型体量，以抵御主入口处强风的侵袭。同时，设计师创造性地在建筑立面上设置波浪形遮阳百叶，不仅呼应了周围碧波粼粼的水面，又与流线型的建筑形态相得益彰，还可遮挡夏日火热的阳光。

Profile

This mixed-use development contains accommodation for a 5-star hotel, international grade-A offices and top commercial facilities. It will be one of the first buildings to be built as part of the large masterplan of the new special economic zone of Hengqin Island District.

Design Concept

Designers aim to provide an eye-catching and modern design which specifically responds to the site, the environment, and Huarong's design aspirations of creating a landmark tower. The organic towers of this project are carefully arranged to provide hotel guests and office users with stunning views of significant landmarks in Macau. Designers get inspiration from the Eight Trigrams to create two linear architectural volumes that stand oppositely. Rotated orientation of the two buildings creates an open space in between, which ensures open views of the buildings. The relationship between buildings and surroundings has been taken into account, respecting the development potential and view corridors of adjacent sites.

In consideration of the climate of coastal areas, the buildings are envisaged as special linear volumes, protecting its main entrances from wind. The façade consisted of louvers in wavy forms shade unwanted glare and accentuate the concept of a rippling water surface.

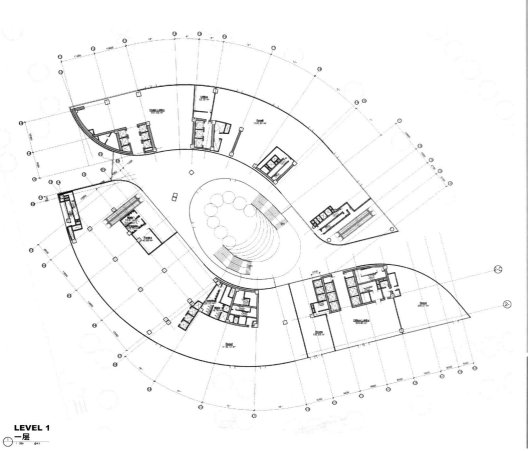

LEVEL 1
一层

Perspective View 1
透视图 1

Perspective View 2
透视图 2

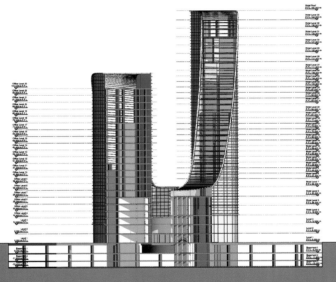
Section 1 1:400
剖面 1 1:400

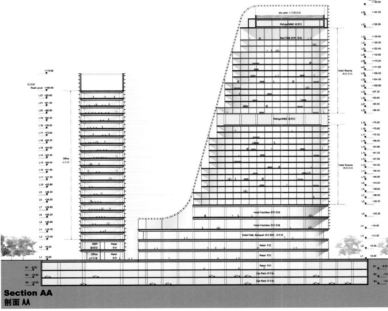
Section AA
剖面 AA

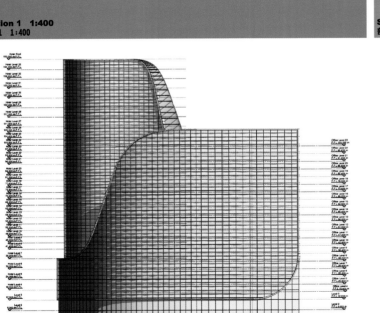

Sectional Perspective
剖面透视图

商业建筑 Commercial Building

"积木块" 超级城墙 山谷空间
"Building Blocks", Urban Super Wall, Valley Space

广东珠海世邦国际商贸中心
Summer International Retail and Entertainment Center

设计单位：10 DESIGN（拾稼设计）
开发商：世邦集团
项目地址：中国广东省珠海市
占地面积：170 000 ㎡
建筑面积：510 000 ㎡
设计团队：Gordon Affleck　Jamie Webb
　　　　　Bernice Kwok　Kevis Wong
　　　　　Nick Chan　Jason Easter
　　　　　Christian Dierckxsens　Ewa Koter
　　　　　Alicia Johannesen

Designed by: 10 DESIGN
Client: Summer Industrial Group
Location: Zhuhai, Guangdong, China
Site Area: 170,000 m²
Gross Floor Area: 510,000 m²
Design Team: Gordon Affleck, Jamie Webb,
Bernice Kwok, Kevis Wong, Nick Chan, Jason Easter,
Christian Dierckxsens, Ewa Koter, Alicia Johannesen

项目概况

坐落于珠海的世邦国际商贸中心是一个包括360 000 ㎡的零售空间、办公区、酒店式服务公寓、住宅区等功能区的综合用途项目，它将是全球最大的零售及综合开发项目之一。

设计构思

项目既处于城市主干道汇聚之处，也处于周边自然山脉的交会之处，以此为设计灵感，以仿"积木块"的形式，将建筑体量沿着围墙区域分散布局，结合穿插在自然景观花园之中的各式积木墙，共同构筑了一座极具动感的"超级城墙"。设计师以极具创意、突破传统的设计构思，将零售建筑从一般的"黑盒子"中解放出来，营造了丰富多样的零售和娱乐空间环境。

设计特色

巨型的石级墙或不锈钢及LED发光积木块，构成了穿越街道的开敞空间及悬挑结构，也界定了项目的边界和城市框架。LED发光积木块既是建筑门户的通道与泛光照明工具，也是"超级城墙"的分隔点。

巨墙之间的空隙，透露出绿化台阶形的建筑和街道，形成一座巨型的神秘花园。设计师沿着起伏的"台阶"建筑"山谷空间"，或将空间打开形成开放式广场，或将空间收缩形成私密的阴凉庭院，为市民提供了多样、非凡的市民空间和公共开放空间。

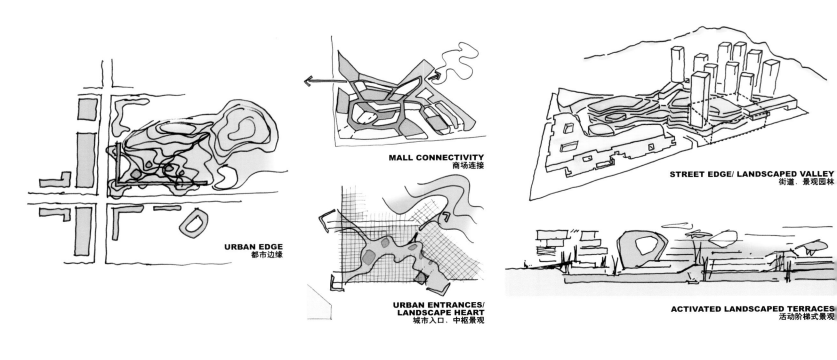

URBAN EDGE
都市边缘

MALL CONNECTIVITY
商场连接

URBAN ENTRANCES/ LANDSCAPE HEART
城市入口．中枢景观

STREET EDGE/ LANDSCAPED VALLEY
街道．景观园林

ACTIVATED LANDSCAPED TERRACES
活动阶梯式景观

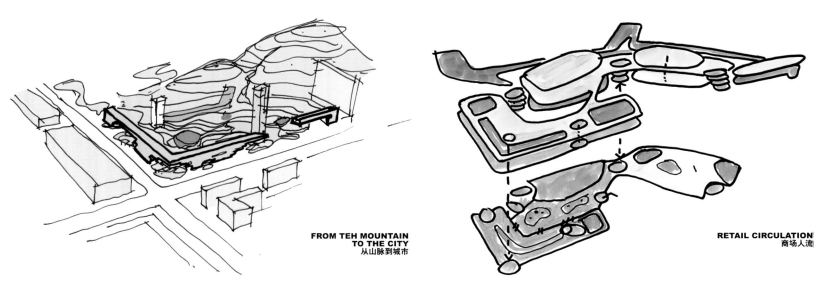

FROM TEH MOUNTAIN TO THE CITY
从山脉到城市

RETAIL CIRCULATION
商场人流

Profile

The mixed use development in Zhuhai, China contains 60,000 square meters leasable retail space together with commercial, hotel serviced apartment and residential accommodation. It will become one of the world's largest retail developments.

Design Concept

The site of the development locates at the meeting point between the grid of the city and the natural topography of the surrounding hill range. The design of the development takes inspiration from this with a dynamic "urban super wall" defining the sites edge with the urban grid. Taking "Building Blocks" as form, building volumes scatter along the wall. Designers with innovative unconventional design ideas free the retail building from the standard "black box", creating a series of rich and diverse retail and entertainment environments.

Design Feature

The super wall is made up of series of giant, stacked stone, steel and LED blocks that are stacked to open and cantilever out across the street as well as define the project boundary and urban framework. The LED blocks acts as media and light entry gateways and break out points.

These breaks gates within the wall reveal the planted terrace building and street within like a giant secret garden. The terraces within the heart of the site are sculpted to reflect the flow of pedestrian movement through the site along undulating terraced valleys that open to create external plazas and close to create intimate shaded courtyards.

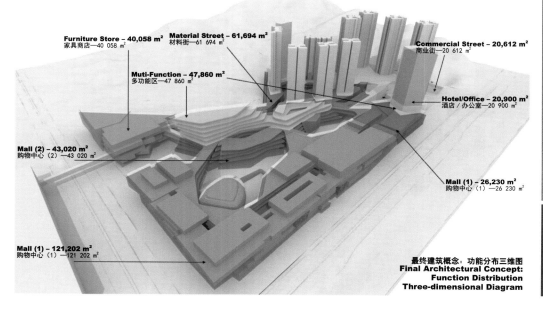

最终建筑概念：功能分布三维图
Final Architectural Concept: Function Distribution Three-dimensional Diagram

土耳其布尔萨蔬果和鱼类批发市场
Bursa Wholesale Greengrocers and Fishmongers Market

设计单位：Tuncer Cakmakli Architects
开发单位：布尔萨市政当局
项目地址：土耳其布尔萨
用地面积：304 000 ㎡

Designed by: Tuncer Cakmakli Architects
Client: Municipality of Bursa
Location: Bursa, Turkey
Land Area: 304,000 m²

体育馆
中亚传统
环线交通

Stadium,
Central Asian
Tradition,
Loop Traffic

设计构思

布尔萨位于土耳其 Uludağ 山脉的山脚，周围土地肥沃，是土耳其最好的农产品生产基地。为此，设计旨在在当地建立一个现代化的商业设施，在完善水果、蔬菜交易空间布局的同时，为鱼类和其他海产品设立单独的交易场所，使之既可促进当地农产品的批发贸易，又为当地的农产品交易活动营造公平、公正、自主协商的商业氛围，成为监控布尔萨地区食品供给的集中控制点。

设计特色

设计打破了将批发市场等功能区间设置在一些无关紧要、毫无个性特征的仓库里的常态，将布尔萨蔬果和鱼类批发市场设置在布尔萨快速发展的地块，这不仅在功能尺度上为批发市场提供了一种新的发展模式，而且使之融入城市风貌中。

建筑的形态酷似体育馆，它有着高高的拱顶，在形态、功能及建筑内涵上与中亚古老的建筑传统联系起来。设计将流动的椭圆形外观与复杂的机动车、货运和行人交通结合起来，在 350m 长的交易市场内，形成清晰、明确、便捷的交通流线，既提高了交易市场的运作效率，又节约了交易成本。

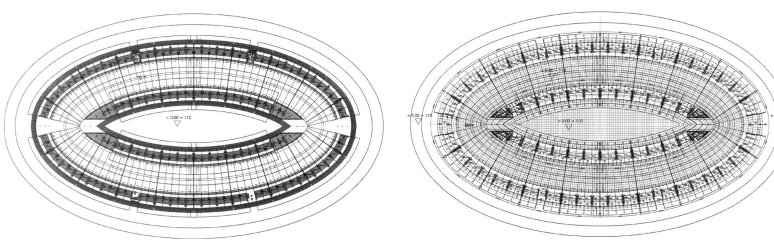

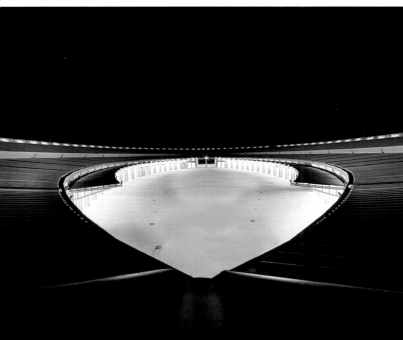

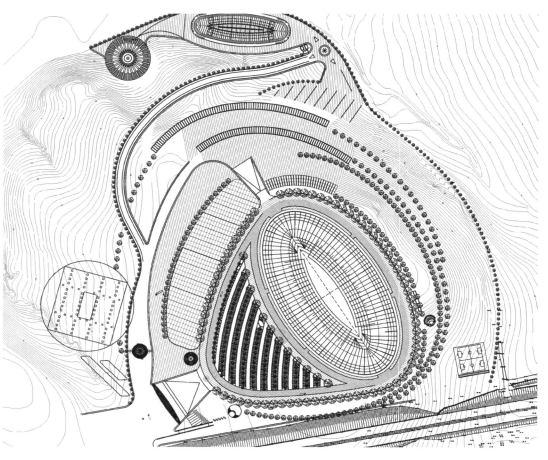

Design Concept

Bursa sits at the base of Uludağ Mountain. The fertile ground around Bursa produce some of the country's best produce. The municipality of Bursa required a modern facility for the wholesale trade of fruits and vegetables, as well as separate facility for fish and other seafoods. The building would consolidate these commercial activities, providing the city with fair, free negotiation commercial atmosphere for local produce trade and a centralized control point from which to monitor the Bursa's food supply.

Design Feature

The project design strives to break a worldwide trend of setting functions such as wholesale trade in architecturally insignificant anonymous warehouse spaces. It locates Bursa Wholesale Greengrocers and Fishmongers Market on a rapidly developing site of Bursa to offer a new development mode and integrate the market with the cityscape.

The forms of the buildings resemble stadiums with high vault, connecting symbolically and functionally with long-standing Central Asian architectural traditions. The complex patterns of vehicle, material, and pedestrian traffic are carefully coordinated within fluid, elliptical shapes. The rational form of the 350 meters long greengrocer's market is designed to facilitate easy orientation, efficient exchange, and optimal routing of foodstuffs from suppliers to retailers and restaurateurs – all of which keeps down transaction costs.

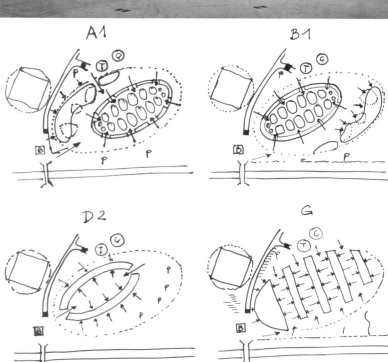

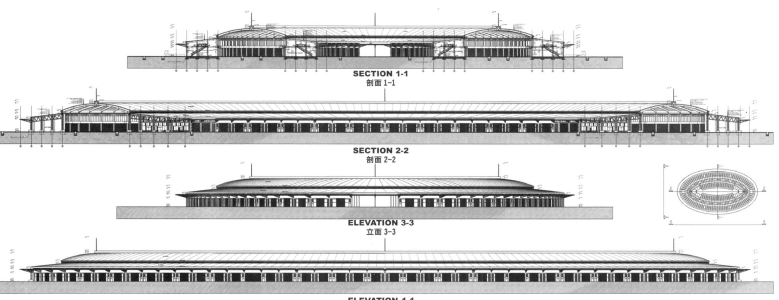

SECTION 1-1
剖面 1-1

SECTION 2-2
剖面 2-2

ELEVATION 3-3
立面 3-3

ELEVATION 4-4
立面 4-4

TRUCK ENTRANCE / PLAN
货车入口 / 平面图

RETAIL ENTRANCE / CROSS ELEVATION
零售入口 / 交叉立面

RETAIL ENTRANCE / CROSS ELEVATION
零售入口 / 交叉立面

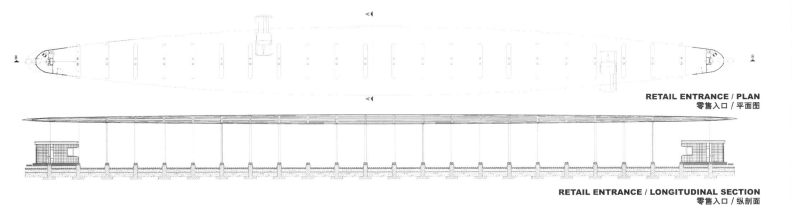

RETAIL ENTRANCE / PLAN
零售入口 / 平面图

RETAIL ENTRANCE / LONGITUDINAL SECTION
零售入口 / 纵剖面

墨西哥合众国墨西哥城佩德雷加尔购物中心
Pedregal Shopping Center

设计单位：Pascal Arquitectos
项目地址：墨西哥合众国墨西哥城
项目面积：7 000 ㎡

Designed by: Pascal Arquitectos
Location: Mexico City, Mexico
Area: 7,000 m²

突破"围城" 生态智能
"Wall" Breaking, Ecological & Intelligent

设计构思
佩德雷加尔位于墨西哥城易被忽略的一个地区，设计旨在在这一地区构建一个极具吸引力的购物中心，不仅满足当地对各种服务设施的需求，同时为该地的发展注入活力。当地的建筑格局好似一座位于石墙内封闭的孤城，"城内"的人不知外面的世界，"城外"的人也不能探知到里面的情况，考虑到这一现状及项目发展需求，设计师试图打破当地原有的建筑格局，使之与外界建立起直接的联系。

立面设计
项目主要通过立面设计来达到设计目标。立面的主要材料是锌板，其上刻有不同形状的镂空。而不同色调的黄色与半透明夹层玻璃的嵌入，宛若室内空间与室外空间之间的透明隔断，既使各空间的活动公开化，又将不同的公共空间融合起来。

交通流线组织
购物中心总面积达7 000 ㎡，包括两层的商业空间、屋顶花园和两层的地下停车场。考虑到购物群体的流动性与出入的便捷性，项目设置了不同的坡道以及特殊的停车空间和电梯。位于第一层地下停车场的汽车送货区，则使后勤服务更加快速便捷。包括屋顶花园在内的花园区设计，为购物提供了休闲空间，丰富了购物体验。

生态智能设计
项目中采用的可持续性设计、节能设计和智能设计也是设计的特色和重点所在。各种自动设施与控制系统的应用，如可自动开启的立面卷帘门和可自我调节的空调系统，体现了建筑的智能性；被动式和主动式的节能设计，则确保了建筑的可持续性和生态性。

Design Concept
This project comes to set a new architectural statement in the Pedregal area of Mexico City which has been neglected. It aims to build an appealing shopping center to provide various service facilities and bring vitality to this area. The way that this building relates with its context, is by breaking up with what is common to the zone, which are big houses in big areas surrounded by very high stone walls which do not let anybody know what is happening outside and vice versa.

Façade Design
This goal is achieved with the main façade that consists of two elements: one of them lining with a zinc plate with large irregular perforations to which a different shades of yellow and translucent laminated glass section is embedded. It allows the view of the interior event from the outside, and at the same time allowing the view of the exterior event from the inside; in such way the public social spaces mix and the limits within the urban and the private become frontiers.

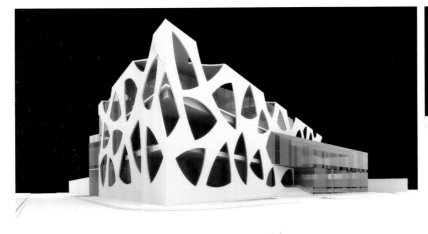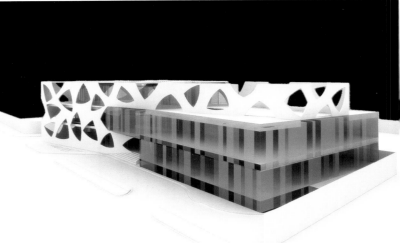

Fachadas

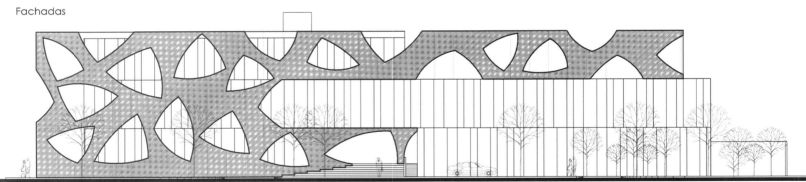

FACHADA PRINCIPAL (AV. FUENTES)

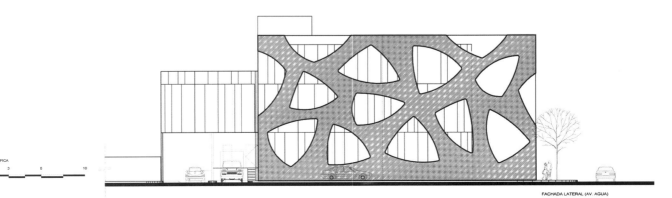

FACHADA LATERAL (AV. AGUA)

ESCALA GRAFICA
0 1 3 6 10
ESC: 1:100

Planta Conjunto

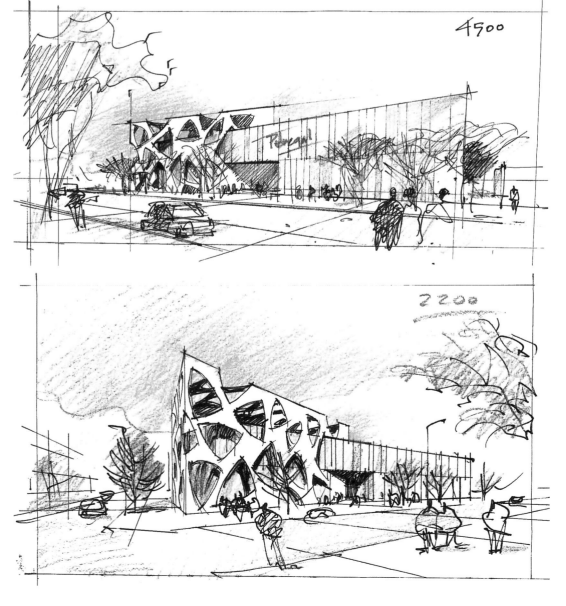

Traffic Circulation Organization

The 7,000 square meter shopping center consists of tw commercial levels, a roof garden and two underground parkin levels. For the consideration of mobility and convenient acces of customers, the project shall set ramps, special parkin spaces, elevators, etc. Numerous garden areas, including th roof offer leisure spaces to enrich shopping experience. Ca delivery zone inside the first parking basement floor ensure fa convenient logistics service.

Ecological Design & Intelligent Design

The project is also characterized and highlighted by sustainable, energy saving intelligent design. Automatio and control systems, such as closing façade rolling door and self-control air conditioning, reflect the intelligence of th building. Passive and active energy saving design ensure th sustainability and ecological features of the building.

Cortes

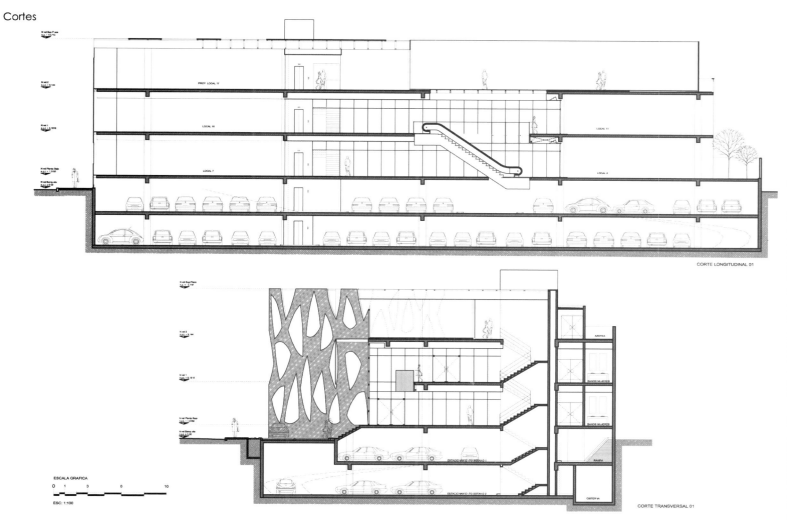

Planta Baja

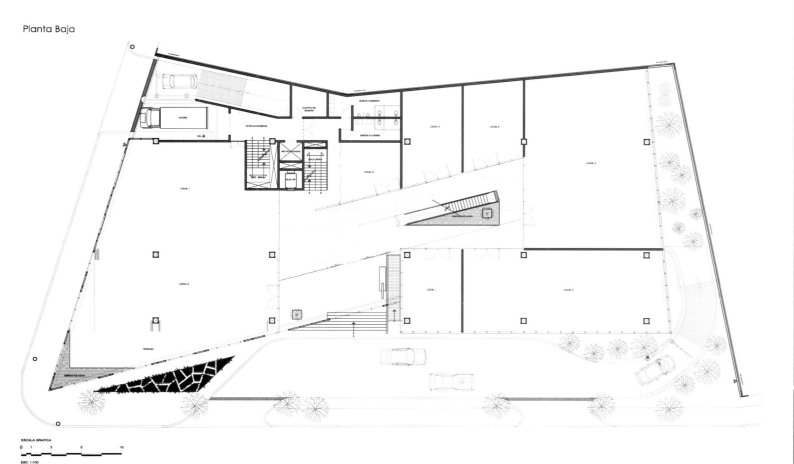

伊朗德黑兰证券交易所
Teheran Stock Exchange

设计单位：LAVA
开发商：德黑兰证券交易所
项目地址：伊朗德黑兰
项目面积：27 380 ㎡

Designed by: LAVA
Client: Teheran Stock Exchange
Location: Teheran, Iran
Area: 27,380 m²

"岩石"
虚拟空间
媒体化立面

"Rock",
Visionary Space,
Media Façades

项目概况

由 LAVA 设计的德黑兰证券交易所融合了未来的几何形式与传统的波斯文化元素，旨在定义新的设计类型，凭借先进的设计工艺，营造一个兼顾视觉性和人的参与性、灵活且可持续的公共中心。

设计特色

当地的传统建筑形式，如洞穴住宅，赋予了设计师极大的灵感与创意。整个建筑体量被抽象为一个巨大的"岩石"，其上分布着精心雕刻的卵形空间，既可以改善建筑的自然采光条件，扩展建筑的景观视野，又可拉近与室外环境的距离，营造宜人的室内环境。

媒体化的外立面宛若显示金融信息的连续数据流，这些数据流沿水平方向漫延开来，将不同层面上的开口空间、突出的构建以及内部空间连接起来，营造一个梦幻、虚拟的实体空间。

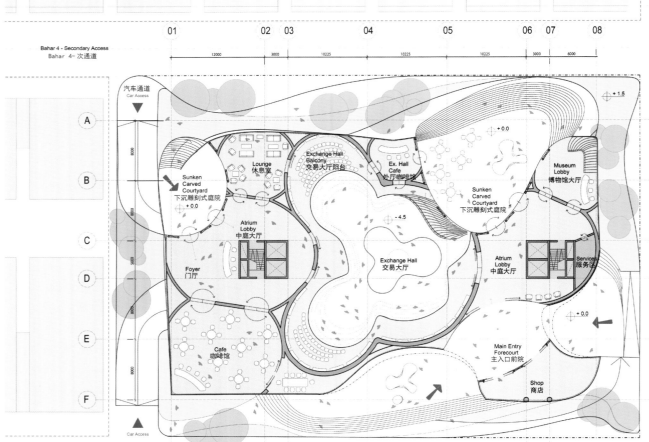

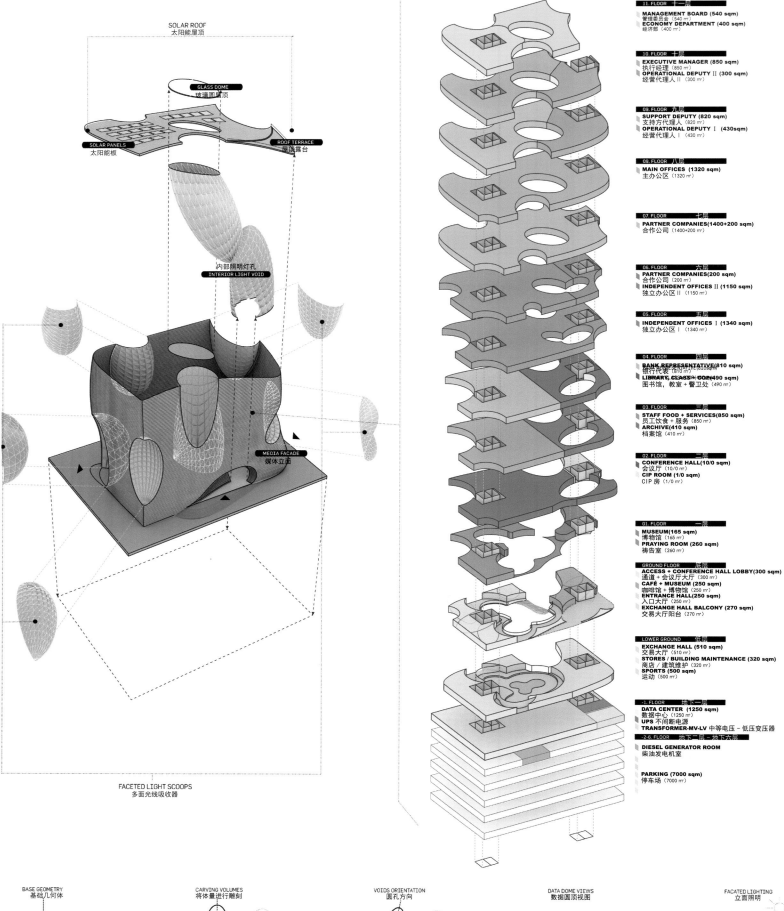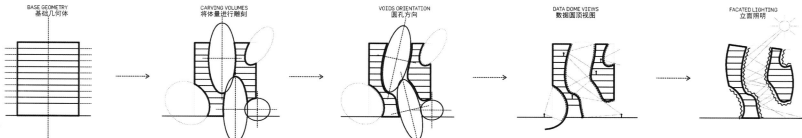

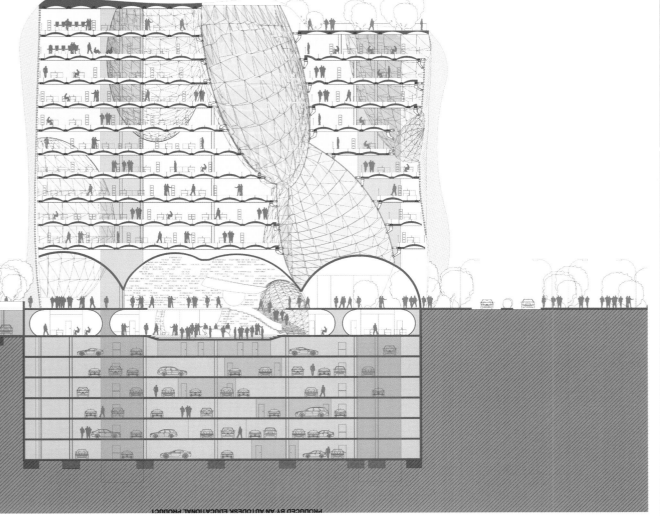

Profile

Teheran Stock Exchange designed by LAVA has fused visionary geometries & forms with traditional elements of Persian culture in the design. The project aims to define a new design type. With most advanced design technologies, the project creates a fully sustainable, flexible and unique public center.

Design Feature

Local traditional architectural forms, such as cave-houses, bring designers great inspiration and originality. The Stock Exchange is envisaged as an urban rock, with smartly carved void shapes that enhance natural light, panoramic views, interior spaces and the relationship with the surroundings. Media façades are a data flow of continuous financial information streaming along a horizontal façade, bent to generate different grades of openness, create carve outs and connection to the interior void, and build a dreamlike visionary space.

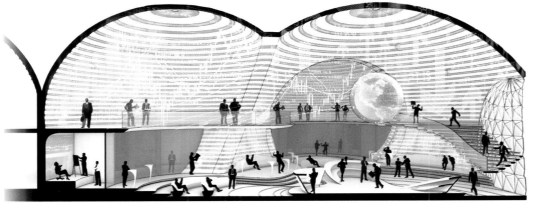

荷兰阿尔梅勒世贸中心
WTC Center Almere

设计单位：de Architekten Cie.	Designed by: de Architekten Cie.
开发商：Eurocommerce B.V., Deventer	Client: Eurocommerce B.V., Deventer
项目地址：荷兰阿尔梅勒	Location: Almere, Nederland
设计团队：Branimir Medić　Pero Puljz　H. de Haas　M. Campschroer　M. Ismael　W. van Meuken　A. Hernandez-Moreno	Design Team: Branimir Medić, Pero Puljz, H. de Haas, M. Campschroer, M. Ismael, W. van Meuken, A. Hernandez-Moreno

项目概况

这一世贸中心综合体是阿尔梅勒市中心重建的一部分，位于中央车站北侧和曼德拉公园之间。

设计特色

整个建筑体由水平方向和垂直方向上的多个体量错列堆叠而成，暗合了该地区整体规划中通过建筑来为阿尔梅勒市树立"瞬间天际线"（Instant Skyline）这一理念。这个由多个建筑体量构成的动态组合，也为建造适合周围环境尺度的建筑提供了参照。

建筑在垂直方向上形成三段式建筑结构：较低楼层的体量进深较大，可以自由访问；中间的楼层部分较为规整；顶部部分则为该建筑提供了室外的全景视野。这样的楼层空间组织形式最大限度地确保了这栋高质量、独具身份特征的建筑的灵活性。

堆叠体量
动态组合
"瞬间天际线"

Staggered Volumes, Dynamic Composition, "Instant Skyline"

Profile

The WTC complex is part of the redevelopment of Almere's city center and is situated between the north side of the Central Station and the Mandela Park.

Design Feature

The design is composed of several volumes that are vertically and horizontally staggered, in response to the masterplan's principle aim of establishing an "Instant Skyline" for Almere through the architecture. This dynamic composition suggests a multitude of buildings within a single complex and at the same time provides a means of tailoring the building's proportions to the surroundings.

The building is vertically segmented into three zones: the lower part of the structure is accessible and deep, the floor area in the central section is average, and the uppermost section offers panoramic views. The generically organized floorspace ensures maximum programmatic flexibility in a high-quality building with a distinctive identity.

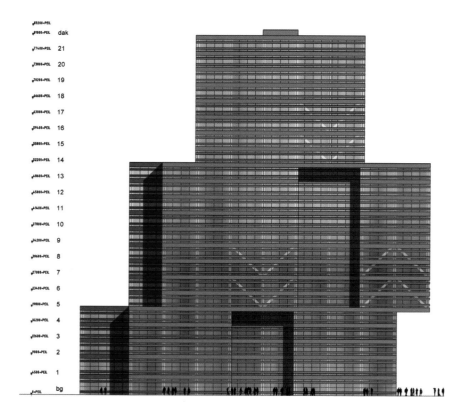
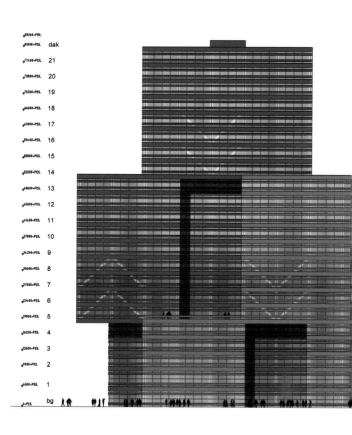

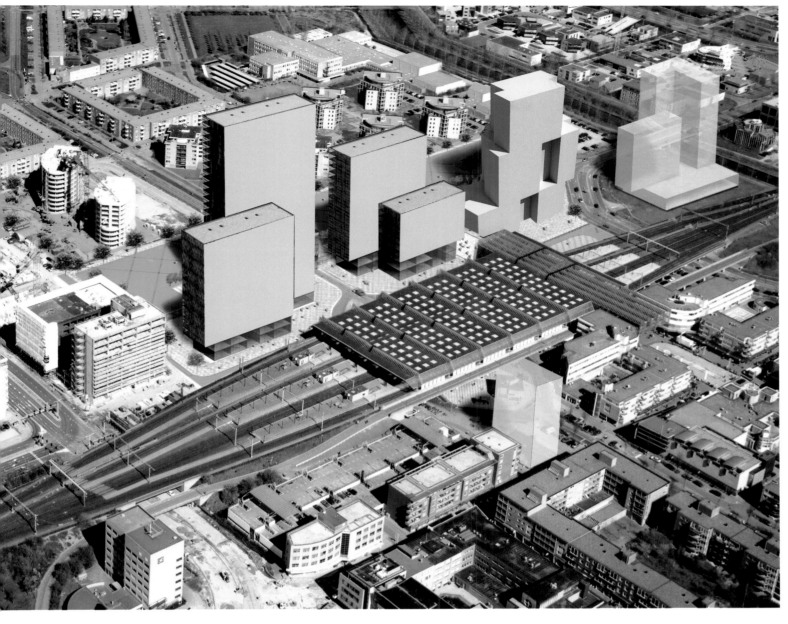
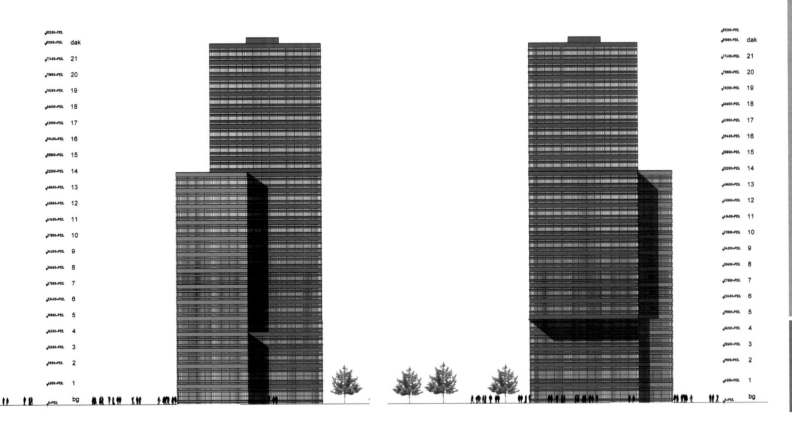

酒店建筑
Hotel

"歌"
流畅节奏
跳跃音符

"Song", Smooth Rhythm, Dancing Music Notes

拉脱维亚里加新 Liesma 酒店
Hotel Liesma

设计单位：绿舍都会
开发商：Liesma 酒店
项目地址：拉脱维亚里加
项目面积：25 000 ㎡

Designed by: **SURE Architecture**
Client: **Hotel Liesma**
Location: **Riga, Latvia**
Area: **25,000 m²**

设计理念

这一方案是绿舍都会对位于拉脱维亚里加海边的 Liesma 酒店进行改造的方案。各学科或领域之间总是存在着千丝万缕的联系，也正因为这种关联性，激发了设计师众多的创意和灵感。在本方案中，设计师将音乐融入建筑中，为这栋老建筑谱写了一曲无声的新"歌"。

设计构思

设计师认为，对新元素和旧元素的理解，不应该是孤立的，而应该站在继承与发展的角度来看待，既建立两者之间不可分割的联系，又能形成新的特征和风格。在本方案中，设计师依据已有的线条来创作新的结构与形式，在已有节奏的基础上，积极探讨水与光、空间与材料、建筑与自然、建筑与景观之间的联系，将建筑的"声音"以及景观的"声音"融入设计中，创作属于自己的"歌"。因此，在这一方案中，新感觉、新灵感空间成为理解本案的唯一方法。

设计特色

不仅仅是建筑曲线的屋顶、建筑外立面上的线条以及表皮上的开口，形成了如跳跃的音符般流畅的动感，建筑的内部空间也传达了一种特殊的节奏，营造了一个新的诗意空间，在视觉上和感官上，给予人们新的认识和体验，让人们不由自主地跟随"歌曲"的节奏调整自己的生活步调。

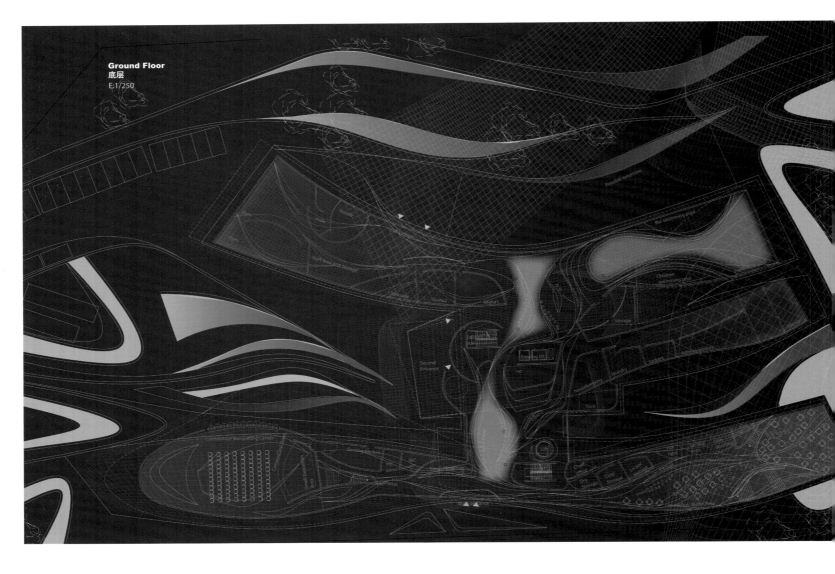

Ground Floor
底层
E:1/250

SKIN LAYOUT
表皮布局

STEEL FRAME STRUCTURE
钢架结构

METAL PANEL
金属板

GLASS PANEL
玻璃板

SECTION AA'
剖面 AA'

SECTION BB'
剖面 BB'

SECTION CC'
剖面 CC'

Design Concept

The scheme is a renovation project designed by SURE architecture for Hotel Liesma in Riga close to the sea. Various subjects and fields are inextricably linked. Such relevance provokes numerous creative ideas and inspiration of designers. In this case, designers bring music into the building, composing a new song for this old building.

Design Concept

Designers believe that new and old elements cannot be separated. From the perspective of inheritance and development, they build indivisible relationship of the new and old elements as well as form new features and styles. In this scheme, designers actively explore the dialogue between water and light, spaces and materials, nature and construction, landscape and building based on the existing rhythm, creating a new structure and form by existing lines. They bring the "sound" of architecture and the "sound" of landscape to the design to compose a "song" of their own. New feelings and new inspiration spaces become the only way to perceive this case.

Design Feature

Curved roof, lines on the façades and openings of the skin create sense of dynamic as dancing music notes. The special rhythm of the internal space creates a new poetic space as well as brings people new recognition and experience visually and perceptually, making them adjust their pace of life along with the rhythm of "song".

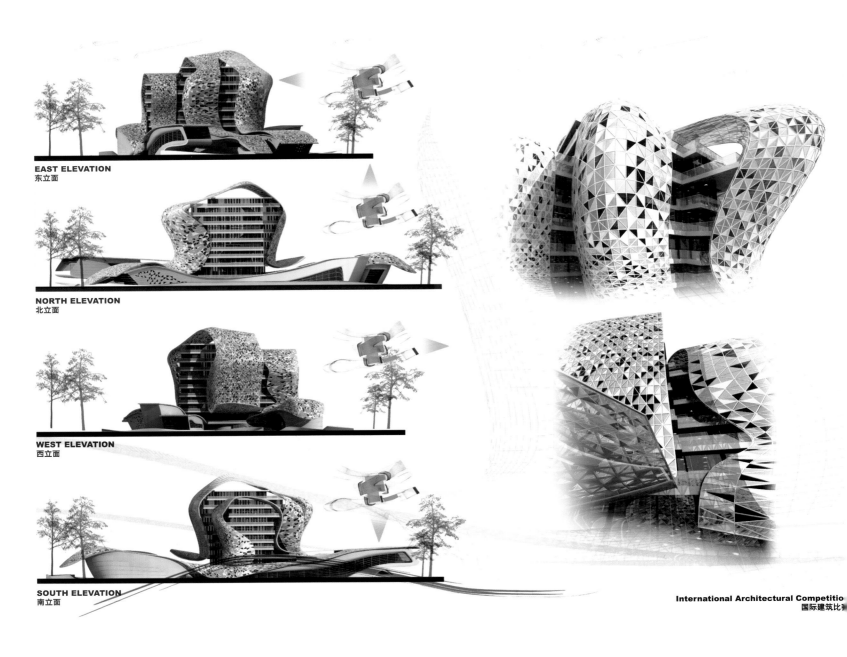

EAST ELEVATION
东立面

NORTH ELEVATION
北立面

WEST ELEVATION
西立面

SOUTH ELEVATION
南立面

International Architectural Competitio
国际建筑比

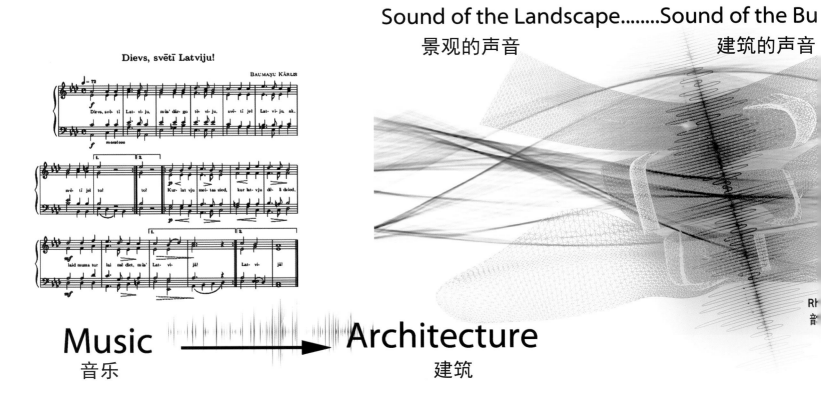

Sound of the Landscape........Sound of the Bu
景观的声音　　　　　　　　　建筑的声音

Dievs, svētī Latviju!
BAUMAŅU KĀRLIS

Music → Architecture
音乐　　　建筑

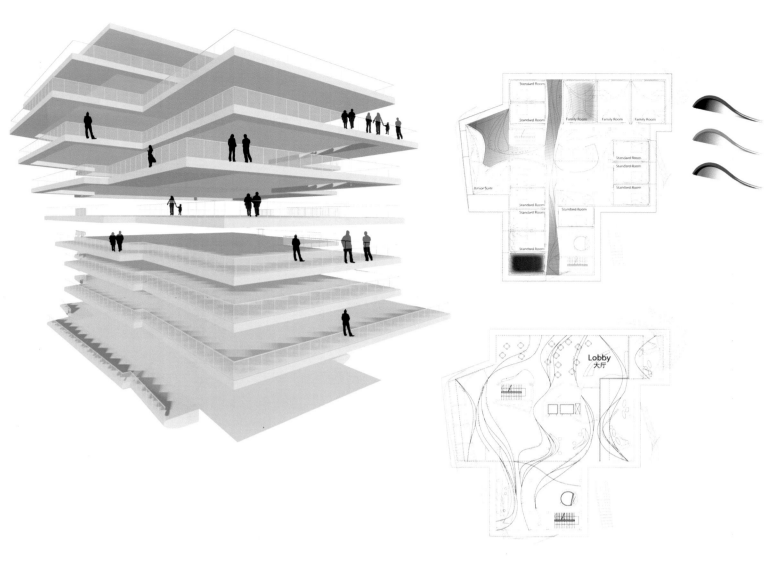

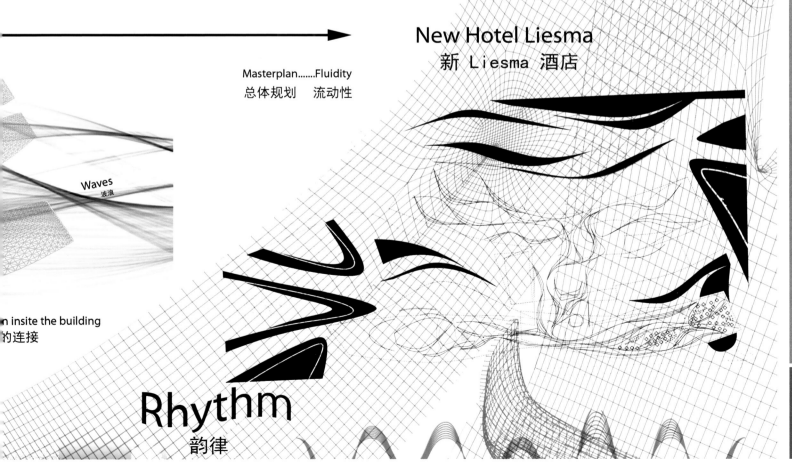

Masterplan……Fluidity
总体规划　流动性

Waves
波浪

n insite the building
的连接

New Hotel Liesma
新 Liesma 酒店

Rhythm
韵律

黑山共和国 Rafailovichi Apart 酒店及地下停车场
Rafailovichi Apart-hotel & Underground Parking

设计单位：TOTEMENT/PAPER	
项目地址：黑山共和国 Rafailovichi	
基地面积：13 800 ㎡	
总建筑面积：7 800 ㎡	
设计团队：Levon Ayrapetov　Valeriya Preobrazhenskaya	
Andrey Panchenko　Diana Grecova	
Julia Presniakova　Andrey Guliaev	
Zoya Nalivaiko　Marina Sipko	
Irina Diakonova　Oleg Burmistrov	
Adelina Rivkina　Evgeni Kostsov	

Designed by: TOTEMENT/PAPER
Location: Rafailovichi, Montenegro
Site Area: 13,800 m²
Total Building Area: 7,800 m²
Design Team: Levon Ayrapetov,
Valeriya Preobrazhenskaya, Andrey Panchenko,
Diana Grecova, Julia Presniakova, Andrey Guliaev,
Zoya Nalivaiko, Marina Sipko, Irina Diakonova,
Oleg Burmistrov, Adelina Rivkina, Evgeni Kostsov

设计构思

曲线，既是平面构图的基本元素，也延续到建筑设计中。设计师将这一项目定义为"蛇"。在相对狭小的建筑场地上，经折叠后的建筑体量被最大化拉长，宛若匍匐在地上蜿蜒前行的蛇，这一曲线的建筑形态实现了对基地的合理利用，也使建筑的各个部分可获得更好的视野和采光条件。

建筑设计

建筑立面由两个相互作用的基本元素组成：黑色玻璃和白色水平装饰层，其位置的变化和流动性让人联想起20世纪六七十年代的图形实验以及当代的动态建筑。项目采用了单边规划结构，别墅从电梯大厅"落下来"，与购物中心和转车平台一起构成后立面的基本元素。

这一地区没有高度的限制，故设计师将场地内的建筑抬高，以创造一个视野区，使位于建筑后方的行人和驾驶员也能享有开阔的视野。停车空间和技术区则尽量隐藏起来。

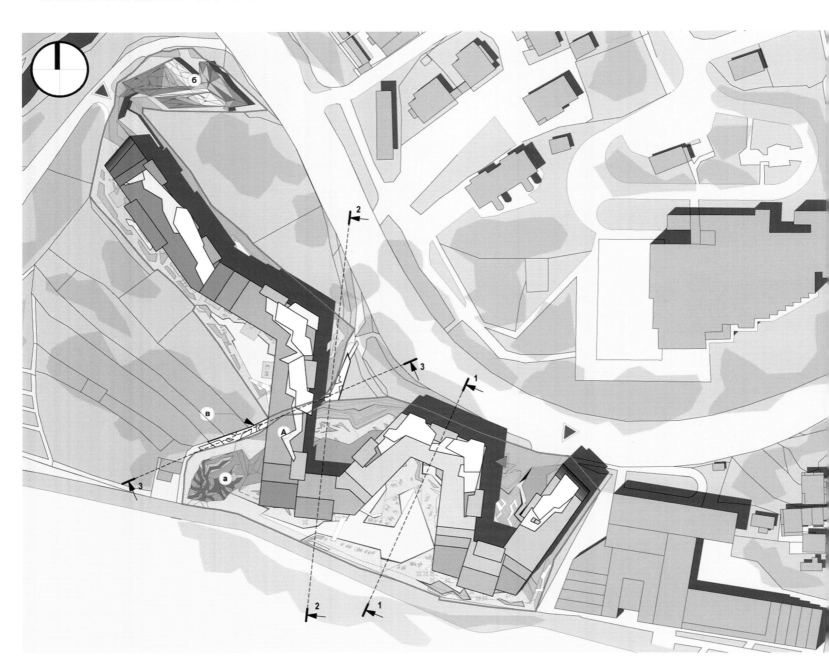

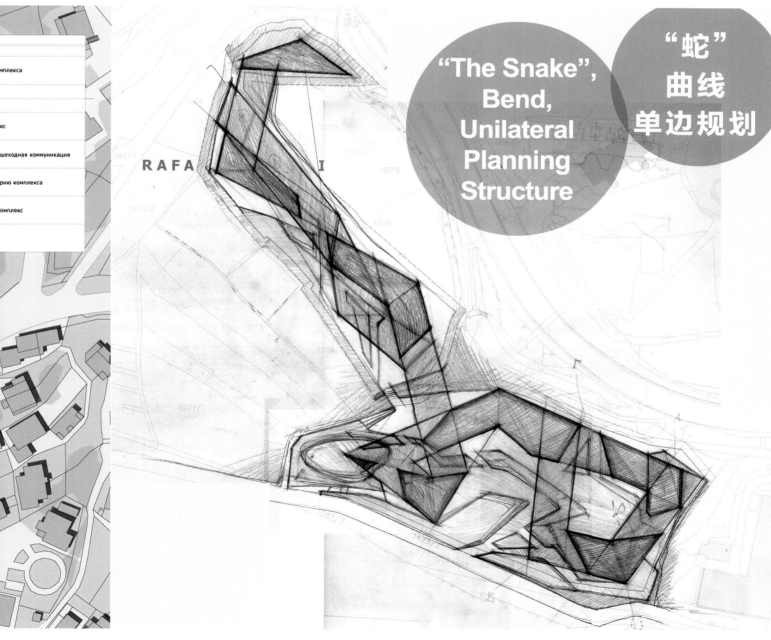

"The Snake", Bend, Unilateral Planning Structure

"蛇" 曲线 单边规划

План 2-го подземного этажа . М 1:1000

План 1-го подземного этажа . М 1:1000

План 5-го (типового) этажа. М 1:1000

План кровли. М 1:1000

Design Concept

Bend, as a basic element of plane, was continued in building design. Designers define this project as "the Snake". On a relatively small platform, the building volume is maximally stretched to occupy the length as much as possible, just as a winding snake. The curved architectural form achieves a rational utilization of the site, and better views and lighting for all cells.

Architectural Design

The façade is made of two interactive elements: a dark glass and white horizontal elements. Their shifting and flowing remind graphic experiments of 60-70's, entered into dynamic architecture recently. Villas "fallen" down elevator halls become the basic elements of the back elevation together with shopping center and turning platform.

In an area lost restriction on height, a building is lifted over the territory to create a zone of viewing, at least for pedestrians and motorists. Parking spaces and technical zones have been hidden in this project.

Зонирование общественных территорий. M 1:1000

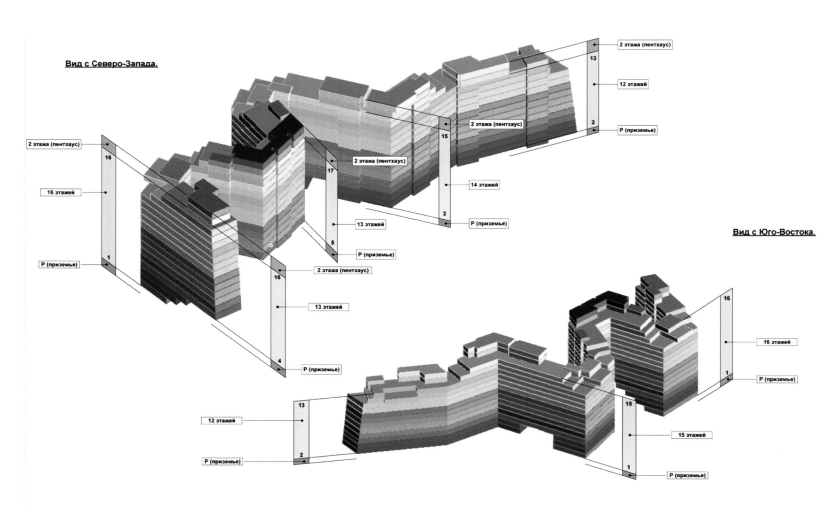

Аксонометрическая схема зоны апартаментов
с указанием этажности.

得标为王——创意篇 2014 | Kings for Bid Winning — Creative Conception 2014

阿联酋阿布扎比埃米尔珍珠酒店
Regent Emirates Pearl

设计单位：Söhne & Partners	Designed by: Söhne & Partners
开发商：Atlas Hospitality, Ahmed S. Al Mutawaa	Client: Atlas Hospitality, Ahmed S. Al Mutawaa
项目地址：阿联酋阿布扎比	Location: Abu Dhabi, UAE
建筑面积：124 000 ㎡	Built Up Area: 124,000 m²
设计团队：Thomas Bärtl　Michael Prodinger　Guido Trampitsch　D. Lems	Design Team: Thomas Bärtl, Michael Prodinger, Guido Trampitsch, D. Lems

"画卷"
多变立面
扶栏阵列

"Picture Scroll",
Varied Façade,
Guardrail Array

项目概况

埃米尔珍珠酒店位于阿布扎比市中心的海滩大道Corniche路上,靠近世界著名的埃米尔宫,将跻身为世界上高端商业酒店的行列。

设计特色

这座饭店的外形令人印象深刻:两座半圆形的饭店附属楼围绕着椭圆形的旋转核心向上盘旋,宛如一卷充满惊喜的神秘画卷,从任何一个角度看都有不同的样貌。建筑底层是一个完全透明的楼层,设置了所有的公共和行政功能。其上5层的墩座通过屋顶与饭店楼层分开,形成一道"裂缝",可在其中建造酒吧、露台、池塘和浴池等设施。顶部设置有直升机停机坪,可以通过直接抵达的电梯前往。

酒店内共设有365间客房,所有的房间都设有阳台,形成一种线性的扶栏阵列。立面上的涂层玻璃板可提供遮阳的功能和开放的视野,金属百叶窗则可调节光线的强弱。饭店里设置了各式餐厅和健身中心,饭店外还有散步码头,形成了综合性的居住环境,为每一位客人提供了舒适的居住体验。

Profile

In close proximity to the world-famous Emirates Palace, at Corniche Road – the beach promenade – a premium location in the center of Abu Dhabi, the planned hotel aims to position and prove itself among high-class business hotels.

Design Feature

The hotel has an expressive form presenting a unique appearance from every perspective. Two semi-circular hotel wings spiral upwards around an elliptical circulation core. The ground floor is a completely transparent volume which contains all public and administrative functions of the hotel. Above it rises a five-story podium volume, which is separated from the towering hotel levels by a second interstice – the Podium Roof – where bar, terrace, pool and jacuzzi are situated. A helicopter pad with direct elevator access is set on the top of the building.

Emirates Pearl Hotel has 365 rooms. All the rooms have balconies which define the building's characteristic outward appearance with a linear composition of railing elements. Coated glass panes provide sun protection and open up views, metal louvers complement these elements. Various restaurants, fitness centers and marina complex provided in this hotel offer comfortable accommodation for every guest.

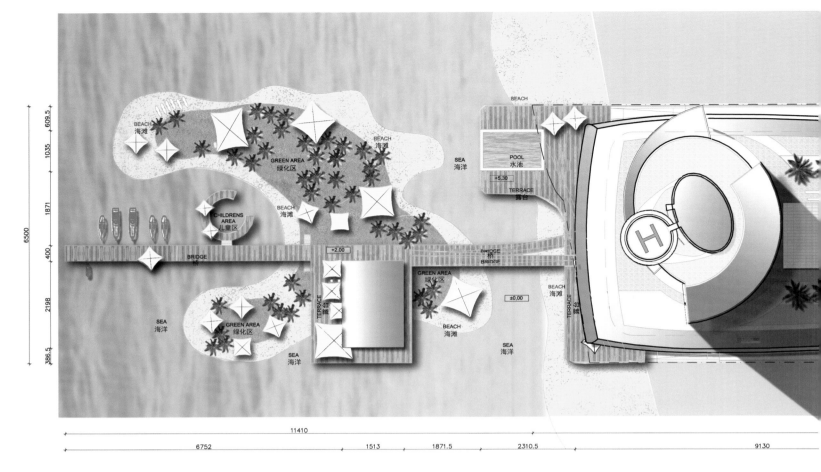

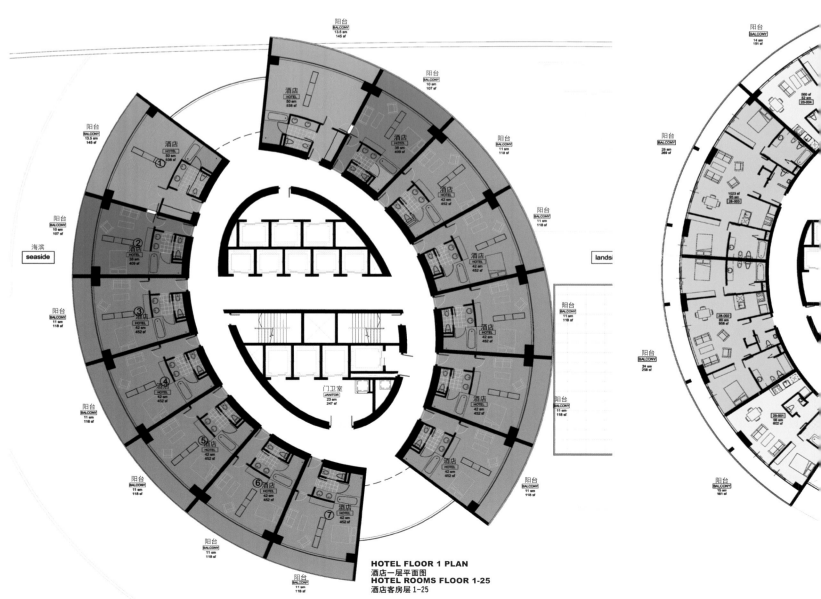

HOTEL FLOOR 1 PLAN
酒店一层平面图
HOTEL ROOMS FLOOR 1-25
酒店客房层 1-25

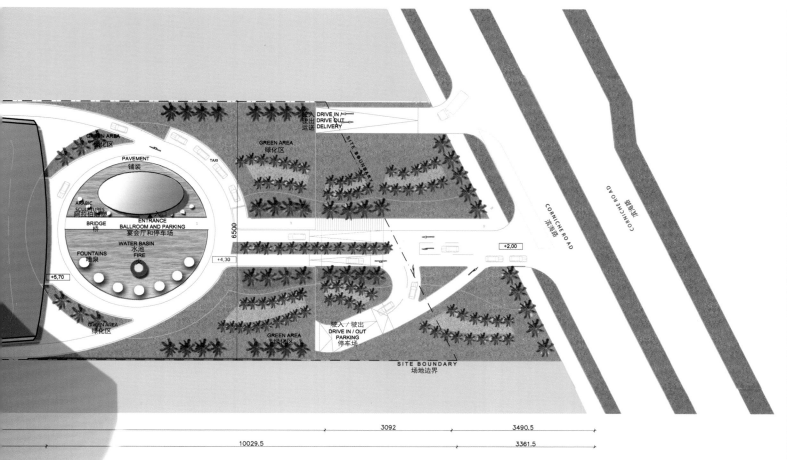

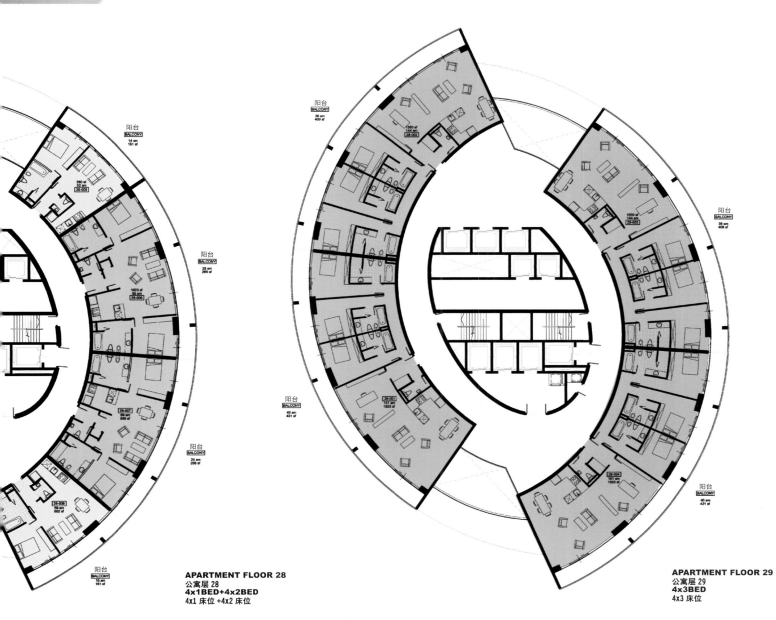

APARTMENT FLOOR 28
公寓层 28
4x1BED+4x2BED
4x1 床位 +4x2 床位

APARTMENT FLOOR 29
公寓层 29
4x3BED
4x3 床位

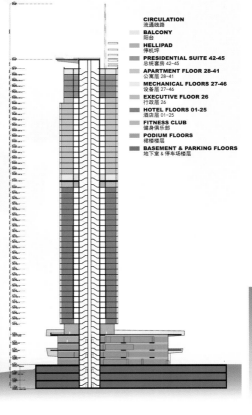
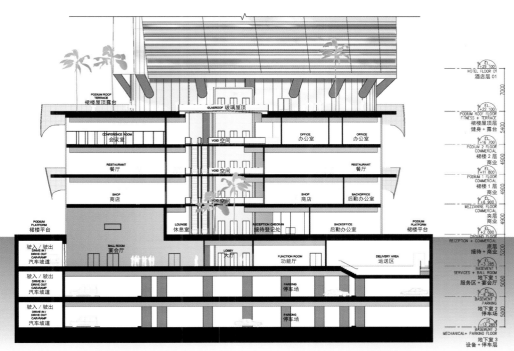

CIRCULATION	流通线路
BALCONY	阳台
HELIPAD	停机坪
PRESIDENTIAL SUITE 42-45	总统套房 42-45
APARTMENT FLOOR 28-41	公寓层 28-41
MECHANICAL FLOORS 27-46	设备层 27-46
EXECUTIVE FLOOR 26	行政层 26
HOTEL FLOORS 01-25	酒店层 01-25
FITNESS CLUB	健身俱乐部
PODIUM FLOORS	裙楼楼层
BASEMENT & PARKING FLOORS	地下室 & 停车场楼层

BALLROOM ABU DHABI CITY HOTEL

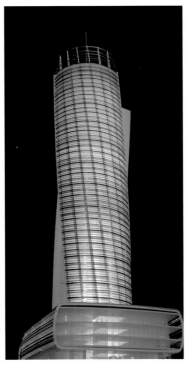

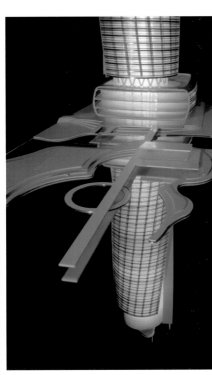

格鲁吉亚巴统 Medea 酒店
Medea Hotel

设计单位：Architetto Michele De Lucchi S.r.l
开发商：Adjara Resort JSC
　　　　Development Solutions LLC
项目地址：格鲁吉亚巴统
场地面积：3 300 ㎡
设计团队：Michele De Lucchi　Leopoldo Freyrie　Marco Pestalozza
　　　　　Marcello Biffi　Laura Parolin　Andrea Saita
　　　　　Sang Yeun Lee
摄影：Gia Chkhatarashvili

Designed by: Architetto Michele De Lucchi S.r.l
Client: Adjara Resort JSC; Development Solutions LLC
Location: Batumi, Georgia
Site Area: 3,300 m²
Project Team: Michele De Lucchi, Leopoldo Freyrie, Marco Pestalozza, Marcello Biffi, Laura Parolin, Andrea Saita, Sang Yeun Lee
Photography: Gia Chkhatarashvili

"闪电" 对角相接 跳舞山峰

"Lightning", Diagonal Connection, Twisting Mountain

项目概况

项目位于巴统的中心地带，本方案是对苏联 Medea 酒店进行的改造方案。优越的地理位置使置身其中的人们可以俯瞰这个古老的城市、海湾以及广阔的大海。

设计构思

现有的建筑是一栋 10/12 层的混凝土结构的体量，这个方正规整、有着狭窄阳台、呈现出社会主义风格的建筑已不能满足当代人的生活需求。考虑到项目优越的地理位置以及传统中心区正交网格的规整布局，设计师没有遵循常规的格栅结构，而是以对角相接的结构，形成强烈的视觉冲击力，给人留下深刻的第一印象。

建筑设计

建筑的形态宛若一道闪电，又好似扭动的山峰，建筑体量的折叠与伸缩，既赋予了建筑不规则的形态，又使整栋建筑从周边规整的建筑群中凸显出来，丰富了城市肌理。整个建筑都被玻璃覆盖着，立面上双网格金属结构的采用，使建筑显得简约而时尚。

入口空间位于朝向大海的主立面处，被一个大型的天棚覆盖，形成了独立的住宅区入口和酒店入口。在这些入口的另一侧，位于底部的 2 层空间被扩大，其中设置了俱乐部和一系列会议室。入口前方的空余空间则被灵活地用作停车空间，以补充停车场有限的停车位。

Profile

Medea Tower is going to be built on the site of the former soviet Medea Hotel. The area is located in the center of Batumi in a prominent position overlooking the old town, the bay and the open sea.

Design Concept

It was originally a 10/12 floor concrete building with a rigorous orthogonal shape in a pure socialistic style with narrow balconies along the two main large façades. It has failed to satisfy life demands of modern people. For the consideration of the project's superior location and the orthogonal grid in the center of Batumi, designers choose to diagonally translate the new building instead of follow the conventional grid, producing a visual impact immediately from the first sight.

Architectural Design

The building form looks like a flash of lightning, a twisting mountain. The folding and stretching of the building volume give the building irregular shape while make the building stand out from surrounding regular buildings. The whole building is covered by glass faced with metal structure using a double grid, which tends to be simple and fashionable.

The entrance is situated on the main façade looking towards the sea and it is covered by a large canopy: entrances to apartments and hotel are separated. On the other side of the entrances the building has an enlargement for the first two floors that host a casino and meeting-rooms. The space left free by the construction is sufficient for a limited number of parking lots in front of the entrance at the sides.

浙江杭州西溪度假酒店

Hangzhou Xixi Tourism Resort Hotel

设计单位：Cervera & Pioz Arquitectos	*Designed by: Cervera & Pioz*
合作单位：华东建筑设计研究院 Belt Collins International Singapore Pte. LTD	*Collaboration: East China Architectural Design & Research Institute Co., Ltd.* *Belt Collins International Singapore Pte. LTD*
开发商：杭州西溪投资发展有限公司	*Owner: Hangzhou Westbrook Investment Co., Ltd.*
项目地址：中国浙江省杭州市	*Location: Hangzhou, Zhejiang, China*
基地面积：3 304.1 ㎡	*Site Area: 3,304.1 m²*
建筑面积：1 329.2 ㎡	*Built Up Area: 1,329.2 m²*
设计团队：José Enrique García Domínguez Rosana Casas Sansó Cristina Álvarez Vicente Macarena Casal Bautista	*Design Team: José Enrique García Domínguez,* *Rosana Casas Sansó, Cristina Álvarez Vicente,* *Macarena Casal Bautista*

"茶杯"破土重生 对话自然
"Tea Cup", Grow Up from the Soil, Dialogue with Nature

设计构思

设计参照了自然中的花卉，旨在将建筑如生长在景观中的花卉或树叶一样，自然地融合在周围的环境中。故设计师设想的不是一个单体建筑，而是5个小型的建筑群，它们由走廊相互连接，就好像花朵或果实由枝干相连一样。

杭州盛产茶叶，中国十大名茶之一——西湖龙井就产于此地。在建筑形态上，为迎合当地的茶文化，设计师将5栋小型建筑设计成5个"茶杯"，其尺度、高度皆不相同，组成一个连续、流畅、线条柔和的综合体，与场地形成对话关系。

建筑设计

为将建筑与场地整合在一起，设计师对基地地形进行处理，使之产生些微的波动起伏，呼应了周围绵延起伏的山峰。5个建筑从基地中拔地而起，宛若从土壤中破壳而出的植物。建筑以圆柱体的形态呈现，其上的缺口改变了建筑的横截面，使圆柱体的形态与玻璃平面交叉后形成一个惊人的立体空间。

建筑外观以玻璃材质为主，双层的玻璃表面依据位置及功能的不同而采用透明或半透明的垂直玻璃模板，形成冬暖夏凉的室内小气候。建筑表面还覆盖了一层竹制的保护层，既是对自然的回应，也能够起到遮蔽的作用。

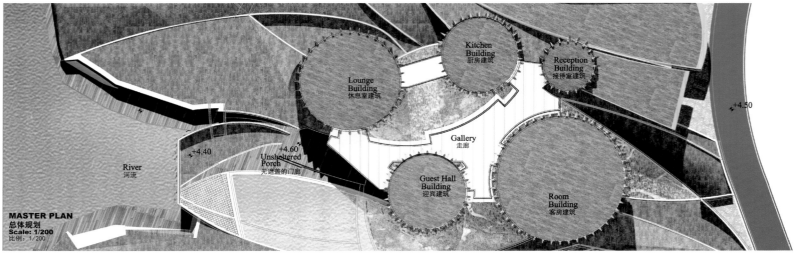

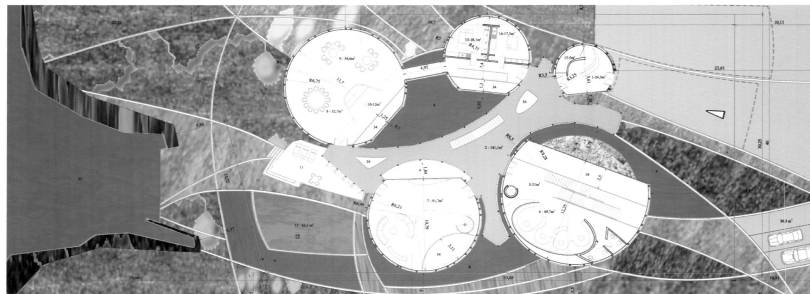

Design Concept

The floral references are the basis of the project. The hotel is integrated in the environment as if it were flowers or leaves laying out in the landscape. Designers envisaged as not just one building but five small pieces connected by a gallery like flowers, seeds or fruits connected by branches.

Hangzhou is the land of tea and one of the top ten famous teas – West Lake Longjing Tea grows here. The project is like a metaphor of this, remembering 5 tea cups. Each one of these small buildings has different size and different height, generating a soft, fluid and kind complex that is able to dialogue with the territory.

Architectural Design

In order to integrate architecture and territory the land of the plot has been transformed by mean of a smooth new topography that moves lightly up and down remembering the hills of Hangzhou. The 5 buildings grow up from the soil as if they were a plant that buds from the land. The forms of the five small pieces or buildings are cylinders, but these cylinders have a cut that transforms the section. This intersection between cylinders and glass leaning planes provokes a surprising spatial perception.

The façade of the five buildings is designed in glass. A façade designed by double layer of glass allows insulation in winter and ventilation in summer. The double glass façade is projected in vertical modules of transparent or translucent glass, according to the different functions inside the building. The façade is protected by an outer layer made in bamboo wood. The wood layer is a poetic echo of the nature whilst shades the five buildings.

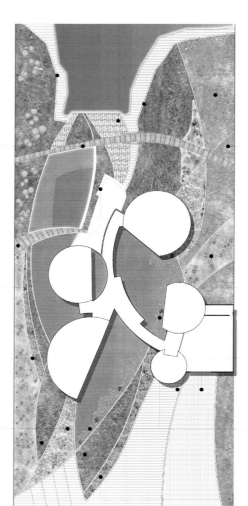

RETAINING WALLS
挡土墙

WOODEN FOOTBRIDGE
木质人行桥

WOODEN STAIRS
木质楼梯

WOODEN FOOTBRIDGE
木质人行桥

TEXTURE OF RETAINING WALLS
挡土墙结构

COVERAGE OF VIOLET AND YELLOW COLOUR FLOWERING
紫色和黄色花卉覆盖区

LAWN OF GRASS TYPE 1
一类草坪

PAVEMENT OF SINGLES NATURAL STONES OVER LAWN
草坪上的自然石铺装

LAWN OF GRASS TYPE 2
二类草坪

BAMBOO FOREST
竹林

NATURAL STONE PAVEMENT
自然石铺装

WATER LILY GROUP
荷花群

COVERAGE OF RED COLOUR FLOWERING
红色花卉覆盖区

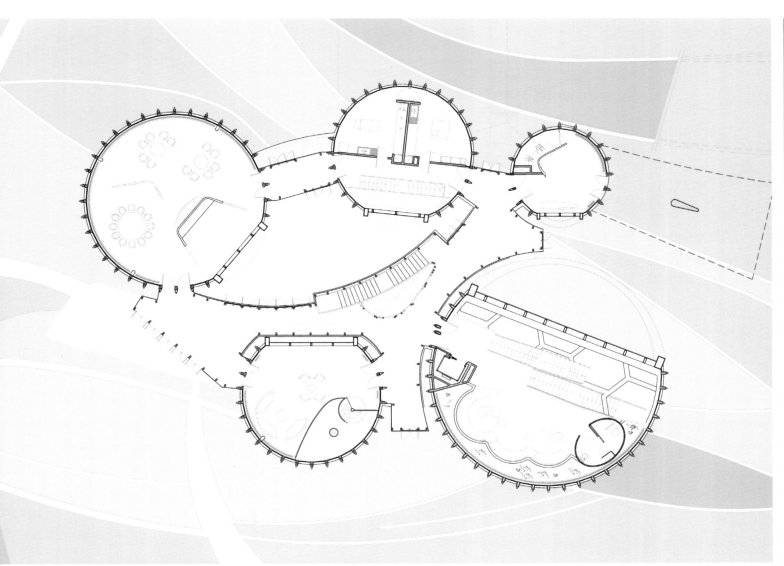

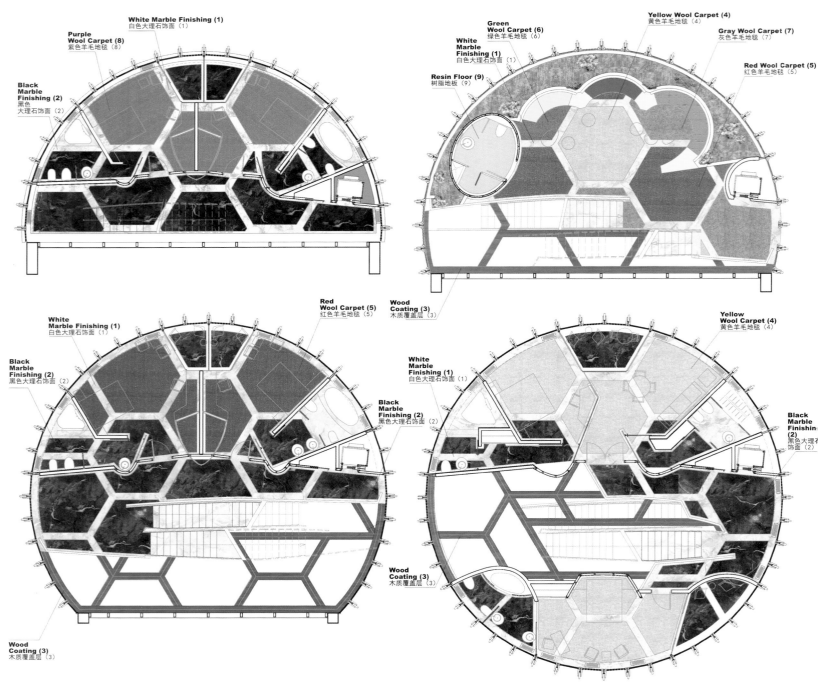

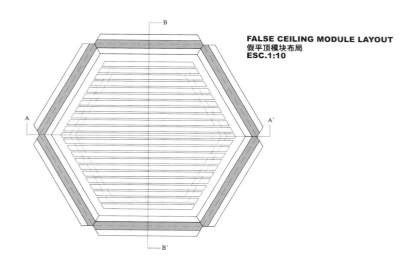

FALSE CEILING MODULE LAYOUT
假平顶模块布局
ESC.1:10

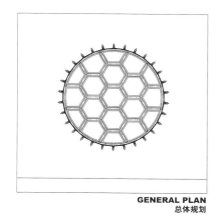

GENERAL PLAN
总体规划

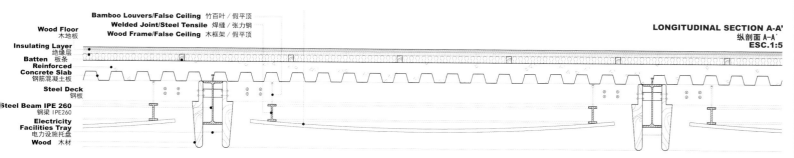

LONGITUDINAL SECTION A-A'
纵剖面 A-A'
ESC.1:5

Wood Floor 木地板
Insulating Layer 绝缘层
Batten 板条
Reinforced Concrete Slab 钢筋混凝土板
Steel Deck 钢板
Steel Beam IPE 260 钢梁 IPE260
Electricity Facilities Tray 电力设施托盘
Wood 木材
Bamboo Louvers/False Ceiling 竹百叶／假平顶
Welded Joint/Steel Tensile 焊缝／张力钢
Wood Frame/False Ceiling 木框架／假平顶

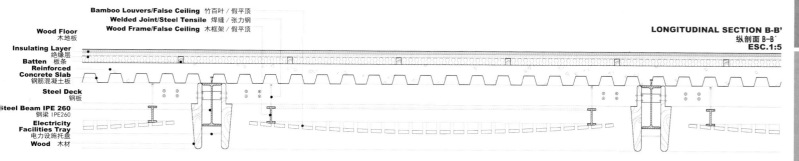

LONGITUDINAL SECTION B-B'
纵剖面 B-B'
ESC.1:5

UNDERGROUND FLOOR FURNITURE PLAN
SCALE: 1:25
地下层家具平面布置图
比例：1:25

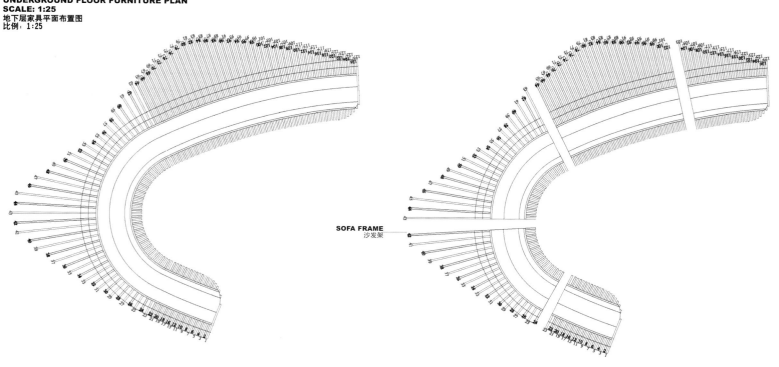

SOFA FRAME
沙发架

NOTE: THE SOFA MUST BE BUILT WITH INDEPENDENT ELEMENTS FOR ASSEMBLY.
备注：沙发必须由独立的元素组合而成。

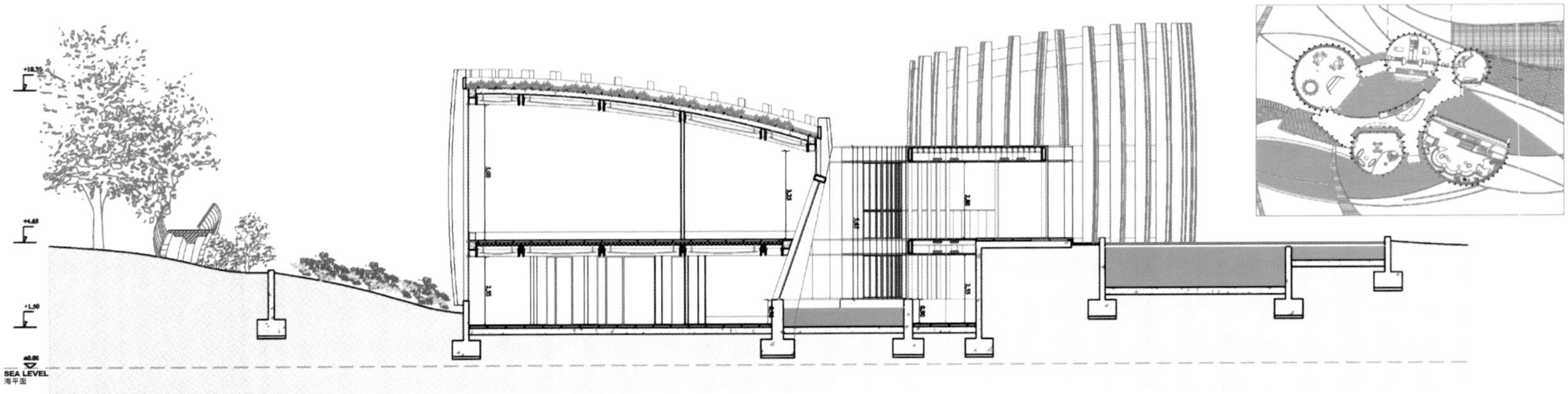

SECTION A-A'
剖面 A-A'

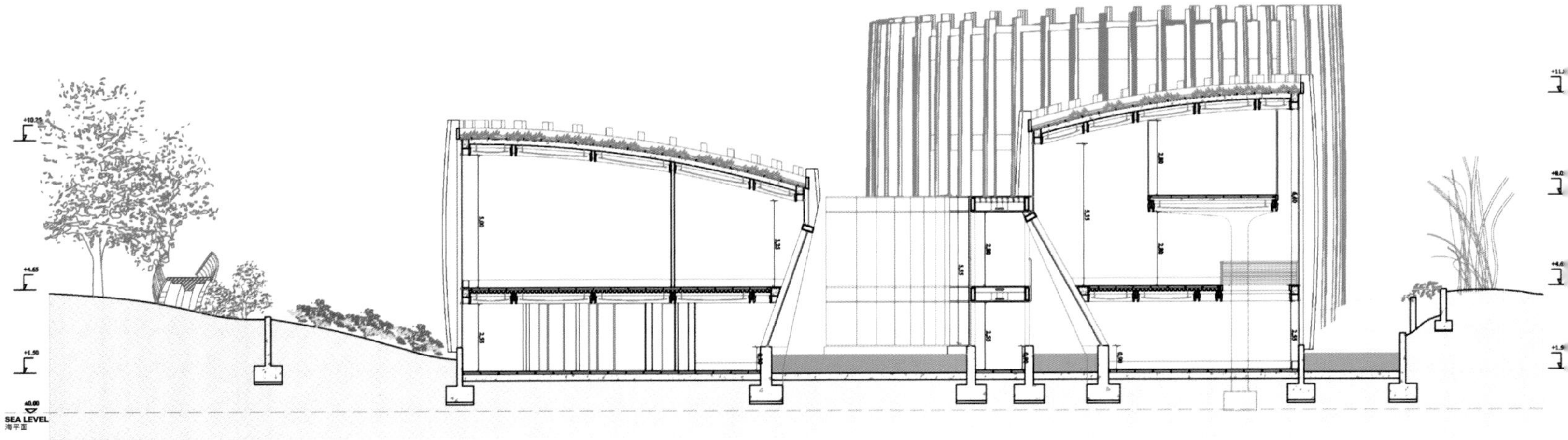

SECTION B-B'
剖面 B-B'

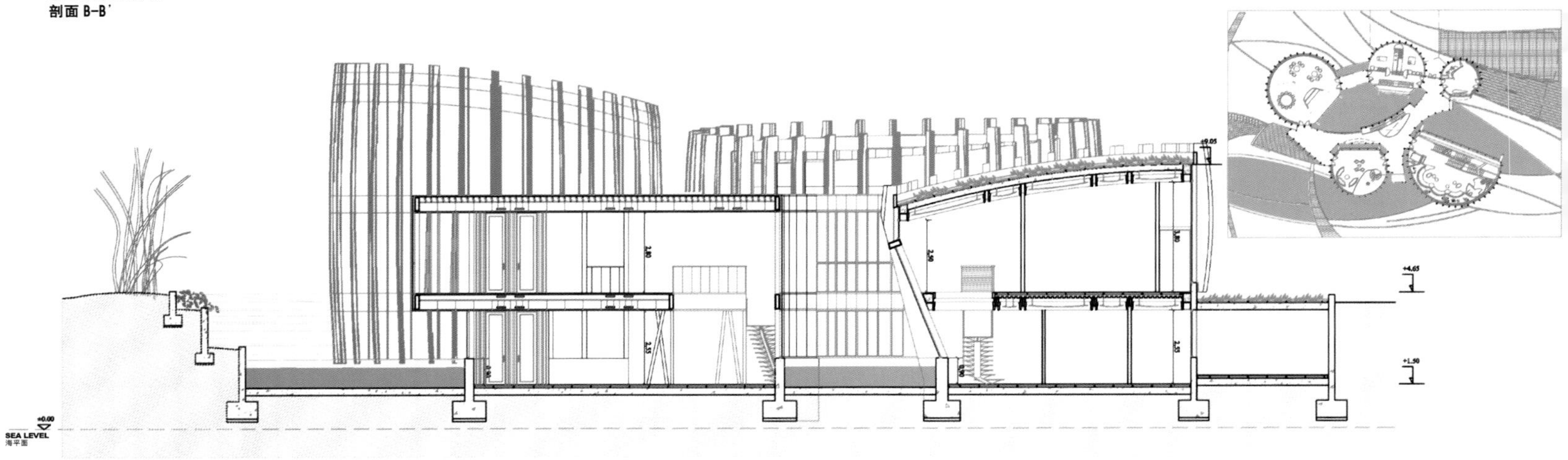

SECTION C-C'
剖面 C-C'

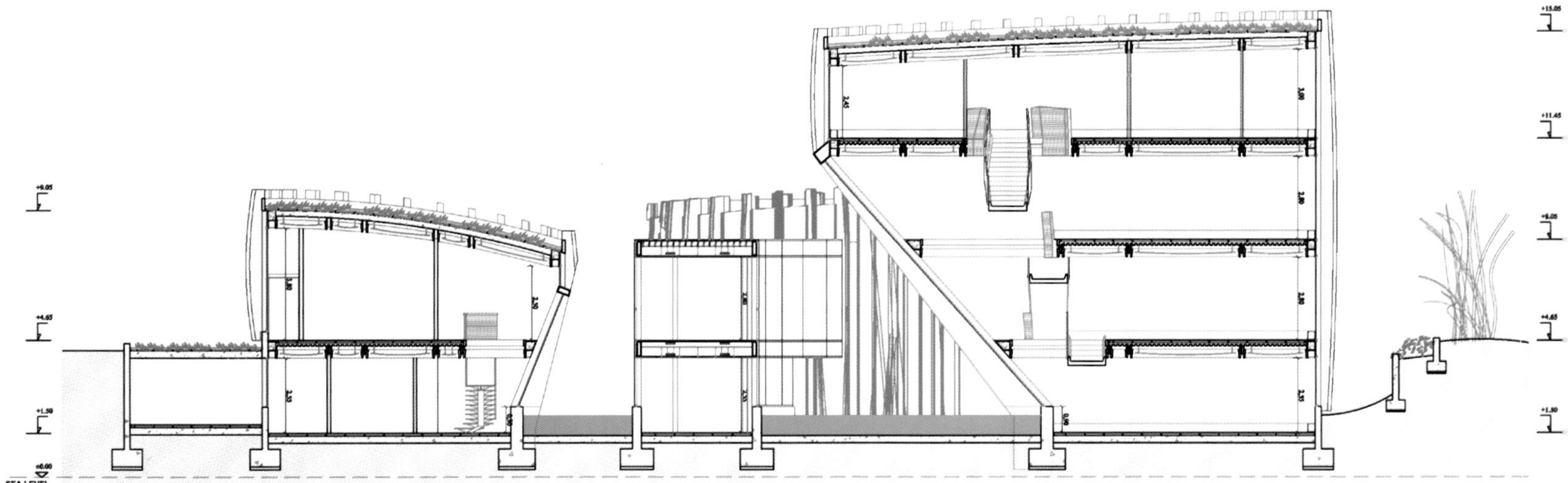

SECTION D-D'
剖面 D-D'

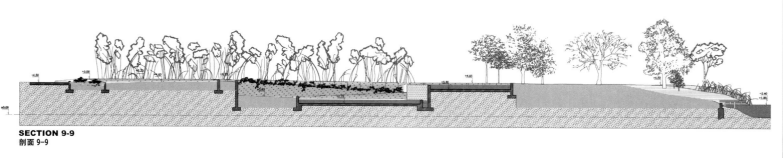
SECTION 9-9
剖面 9-9

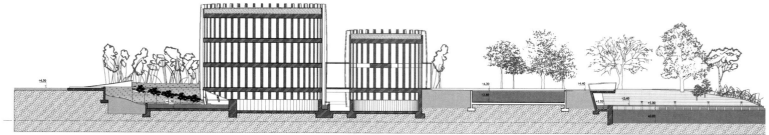
SECTION 10-10
剖面 10-10

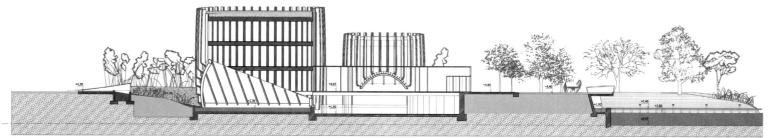
SECTION 11-11
剖面 11-11

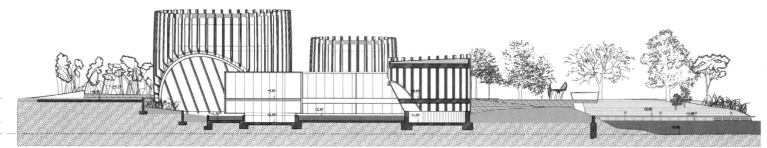
SECTION 12-12
剖面 12-12

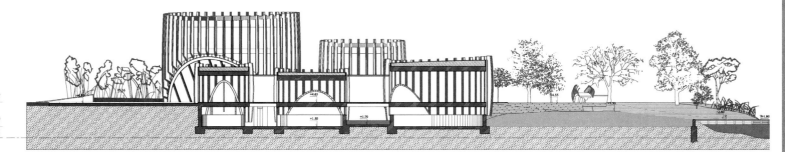
SECTION 13-13
剖面 13-13

SECTION 14-14
剖面 14-14

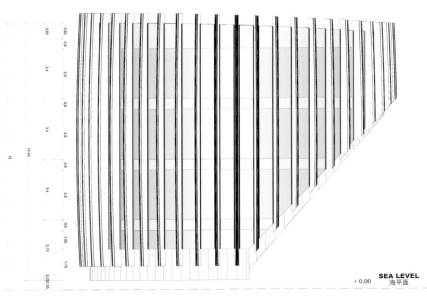

EAST ELEVATION
东立面

SEA LEVEL
+ 0.00   海平面

NORTH ELEVATION
北立面

CONSTRUCTION STEPS. NONE SCALE
施工步骤．不成比例

1. External Building Finishing:
Glazing and Wood Covering on Structural Column.
1．建筑外饰面：玻璃和木材覆盖结构柱。

2. Ball-and-Socket Steel Joints.
To Support Lattice Module and Let Them Rotate by Means of Mechanical System.
2．球窝式钢筋接头。
支撑晶格模块并且使它们依靠机械系统旋转。

3. Lattice Modules. Joined to Ball-and-Socket Joints All Along Straight Supporting Ribs. Close Position during the Night.
3．晶格模块沿笔直的支撑肋连接球窝式接头，夜晚闭合。

4. Lattice Modules. Open Position during the Day.
4．晶格模块，白天打开。

SITE PLAN REFERENCE
总平面参照

CONSTRUCTION SEQUENCE. MATERIALS AND COLOURS CONCEPT.
施工步骤．材料和颜色概念

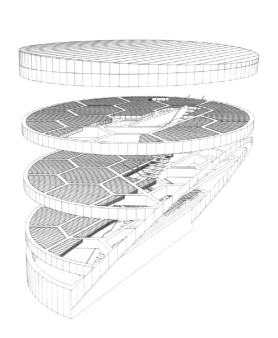

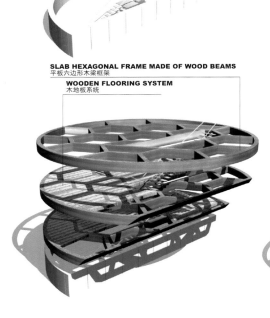

WHITE CONCRETE PODIUM
白色混凝土墩座墙

SLAB HEXAGONAL FRAME MADE OF WOOD BEAMS
平板六边形木梁框架

SLAB HEXAGONAL FRAME MADE OF WOOD BEAMS
平板六边形木梁框架

WOODEN FLOORING SYSTEM
木地板系统

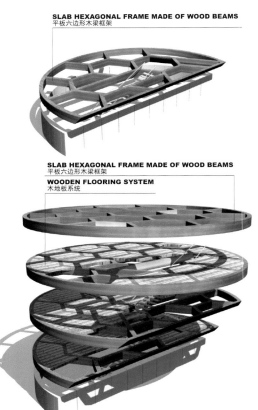

SLAB HEXAGONAL FRAME MADE OF WOOD BEAMS
平板六边形木梁框架

WOODEN FLOORING SYSTEM
木地板系统

CONSTRUCTION SKETCH
建筑示意图

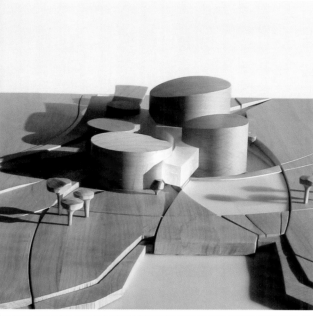

SLAB HEXAGONAL FRAME MADE OF WOOD BEAMS
平板六边形木梁框架

WOODEN FLOORING SYSTEM
木地板系统

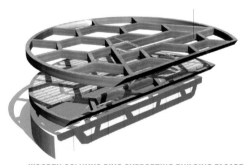
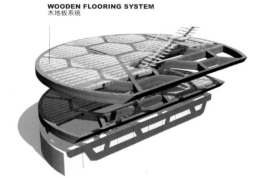

GREEN TERRACE GARDEN
绿色露台花园

WOODEN COLUMNS RING SUPPORTING BUILDING FAÇADE
环状木柱支撑建筑立面

TRANSPARENT AND WHITE TRANSLUCENT WINDOWS DEPENDING ON INSIDE DISTRIBUTION
根据建筑内部布局采用透明窗体和白色半透明窗体

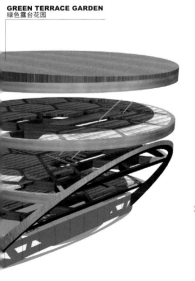

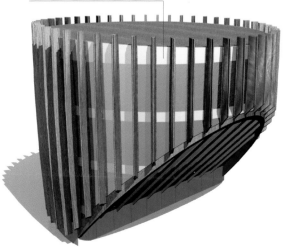

交通建筑
Transportation Building

哈萨克斯坦阿斯塔纳火车站
Astana Railway Station

设计单位：LAVA
项目地址：哈萨克斯坦阿斯塔纳
项目面积：31 000 ㎡

Designed by: LAVA
Location: Astana, Kazakhstan
Area: 31,000 m²

毡房
斜交结构
绝热表皮

Yurt,
Diagrid Structure,
Insulating Skin

项目概况

阿斯塔纳火车站作为该市的三个关键场所之一,不仅是一个交通枢纽,还是一个集社交、商业、贸易、工作、娱乐等功能于一体的活动枢纽,一个彰显城市身份、象征的标志。

设计构思

阿斯塔纳被称为欧亚大陆的心脏,曾是横跨欧亚的古代丝绸之路上的一个重要站点。自古以来就是重要的贸易通道,使"热情好客"成为哈萨克文化至今仍备受推崇的既定传统。故在本方案中,设计师将阿斯塔纳火车站这一国际性的火车站定位为一个独特的会面及聚集场所,建筑形态则参照哈萨克传统民居形式——毡房,以此来体现欧洲中心车站的设计传统以及哈萨克人的热情好客。

设计特色

传统毡房的结构十分注重建筑的功能性,其木质斜交格栅结构外面都会覆盖兽皮保护层,这一特征也被运用到阿斯塔纳火车站的设计中,具体体现在车站的混凝土与钢材改良斜交结构中。这一结构的表面覆盖了一层透明的绝热表皮,是一种与环境相适应的乙烯四氟乙烯聚合物覆盖层,这一表皮不仅对建筑具有防护作用,而且使建筑显得更为轻盈和开放。

Profile

As one of three critical central places within the City, Astana Railway Station will be a central focus for movement, a hub for social, commerce, trade, work, play and other activities, and a symbol showing the identity of the city.

Design Concept

Astana sits astride one of the ancient silk roads and is both a critical junction and a stop within the European-Asian landmass. To meet and greet travelers is an established tradition in Kazakh culture and is honored today. So Astana's International Railway Station should be a place for special meeting and greeting. The project design references the traditional Kazakh house – the yurt to show the traditions of Grand Central stations in Europe and traditional Kazakh hospitality.

Design Feature

The yurt emphasizes the functionality of the building. Its timber diagrid structure is covered with a protective shroud of animal skins. This is reflected in the concrete/steel modified diagrid structure of the Astana Railway Station with a transparent, yet insulating skin – a shroud covering of climatic responsive EFTE. The structure, not only for protection, but also conveys lightness and openness of the building.

"金属树"
广场竹林
棱柱曲面

"Metallic Tree",
Bamboo Forest,
Prismatic Surface

意大利那不勒斯 Garibaldi 广场

Piazza Garibaldi

设计单位：Dominique Perrault Architecture
开发商：Ti Metropolitana di Napoli, Ing. Giannegidio Silva
项目地址：意大利那不勒斯
基地面积：59 000 ㎡
建筑面积：21 000 ㎡

Designed by: Dominique Perrault Architecture
Client: Ti Metropolitana di Napoli, Ing. Giannegidio Silva
Location: Naples, Italy
Site Area: 59,000 m²
Built Area: 21,000 m²

项目概况

Garibaldi 广场是那不勒斯交通体系中最重要、最复杂的交通枢纽之一，这一基础设施包含了一个地铁站，其设计将为提升这一充满活力的城市空间创造条件。

设计特色

这一广场被周围鳞次栉比的建筑群围合而成，形成一个方正、规整、交通流线明晰的公共空间。整个开放空间包括城市公园、绿意盎然的花园、大型的喷泉广场、一个环境保护区、一个覆盖了大面积绿廊的地窖以及两侧林立着精品店的开放型长廊，这些基本元素共同构成了这个热闹、秩序井然而又富有诗意的站前广场。

虽然在结构和选材上，开放型长廊的顶棚与车站不同，却恰到好处地契合了中央车站屋顶的延伸结构。长廊由 8 棵"金属树"构成的钢架结构支撑，每 3 组经过简单变形的"金属树"结构形成一个极具立体感的支撑框架，宛若一片多节、灵活的竹林。其上的覆盖层是一个棱柱曲面，由不同密度和类型的穿孔金属板构成，因此，其外观会不断发生变化。

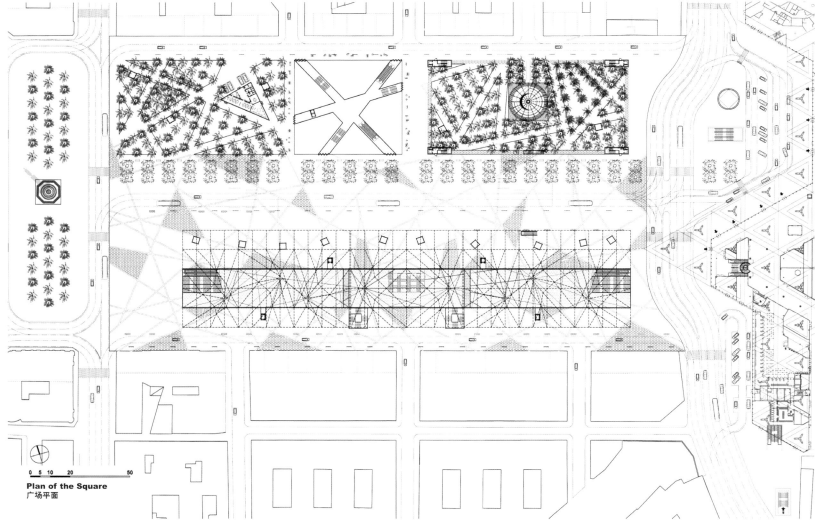

Plan of the Square
广场平面

Profile

The Piazza Garibaldi is one of the most important and complex transportation hubs in the Neapolitan transportation system. This infrastructure project, which includes a metro station, offers the opportunity to upgrade this lively urban space bustling with activity.

Design Feature

Piazza Garibaldi is surrounded by high & low building clusters to create a regular, order public space with clear traffic lines. Open spaces are composed of urban parks, luxuriant gardens, large ponds, a protected area, a hypogeum covered with a large pergola and an open promenade with boutiques lining both sides. These elements jointly constitute a bustling poetic station square.

Though structurally and materially different, the new roof fits right into the alignment and the extension of the central station's roof. A series of eight metallic trees, in simple variations of three patterns, creates a framework resembling clusters of knotty and flexible bamboos. The covering is a vast prismatic surface composed of different types of perforated metal in varying densities, whose appearance is constantly changing.

Metro Access Section
地铁通道剖面

台湾高雄港和游轮中心
Kaohsiung Port and Cruise Center

设计单位：Tonkin Liu
开发单位：高雄港务局
项目地址：中国台湾省高雄市
项目面积：72 648 ㎡

Designed by: Tonkin Liu
Client: Kaohsiung Harbour Bureau
Location: Kaohsiung, Taiwan, China
Area: 72,648 m²

Sea of Flowers, Coastal Forest

梦幻花海 海岸森林

设计特色

远远望去，整个建筑好似笼罩在一片花海中，也像是处在成片的保护伞下。这一设想源自自然中绽放的花朵，当众多花瓣依据固有的几何形态呈一定逻辑拼凑起来，则可形成一道完美的屏障。弧形的"花瓣"铺展开来，一片紧挨着一片，填补了原有的缝隙，形成一个无缝隙的顶棚，花茎则成了空间的支柱，支撑起层叠的"花瓣"。"花瓣"上大小不一的圆形孔洞，既似花瓣上的露珠，又像雨伞上的雨滴，赋予了建筑一种诗意的美感与想象。

这一标志性建筑的设计方案展示了台湾的自然属性，这不仅体现在建筑形态与结构上，也表现在建筑与景观的关系上。项目构建在港口边缘，沿着海岸线分布。为了使建筑的视野更为开阔，设计师将建筑抬高，通过级级抬升的阶梯使之与游轮处于同一高度。逐级抬升的阶梯将人们带到顶层平台，周围的玻璃表面模糊了室内外的空间界限，人们可以自由地欣赏窗外的天空、海洋以及来来往往的船只。

绿色设计

"森林"结构的顶棚在过滤太阳光的同时，也有助于自然通风，营造一个凉爽的室内空间；收集到的雨水经过处理后，可作为建筑用水；在最热的季节，这一建筑还可利用海水制冷，这一系列的措施，在很大程度上降低了建筑的能耗。

Design Feature

Seen from afar, the whole building looks like covered by a sea of flowers, or a large protective umbrella. The inspiration comes from blooming flowers of the nature. Numerous petals are logically put together in certain geometrical form to create a perfect barrier. Arc-shaped "Petals" are spread but connected with each other to fill up the existing gaps and make a seamless canopy. The stems become braces of the space to support stacked "Petals". Round holes of different sizes on the "Petals" resemble dewdrops on petals or raindrops on umbrella, giving the building poetic aesthetics and imagination.

The landmark building celebrates the nature of Taiwan. It is not only reflected on the architectural form and structure but the relationship between architecture and landscape. The project is built at the edge of the port and distributed along coastline. To make an open view, designers raise the building to the level of the liners by steps. The steps lead people to the top platform. The gazed surface blurs the boundary between the indoor and outdoor spaces, which enable people to freely enjoy the view of the sky, the sea and ships.

Green Design

The structural forest filters the sunlight and facilitates natural ventilation, creating a cool interior space; the collected rainwater after treatment can satisfy water needs of the building; in the hottest months, the building can use seawater to achieve additional cooling, which to a large extent reduces energy consumption.

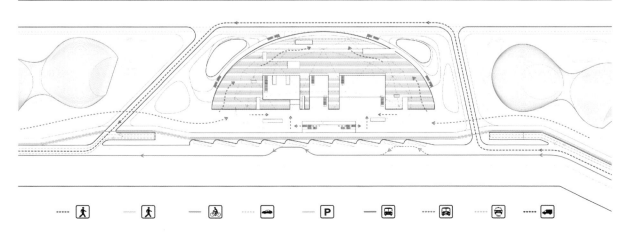

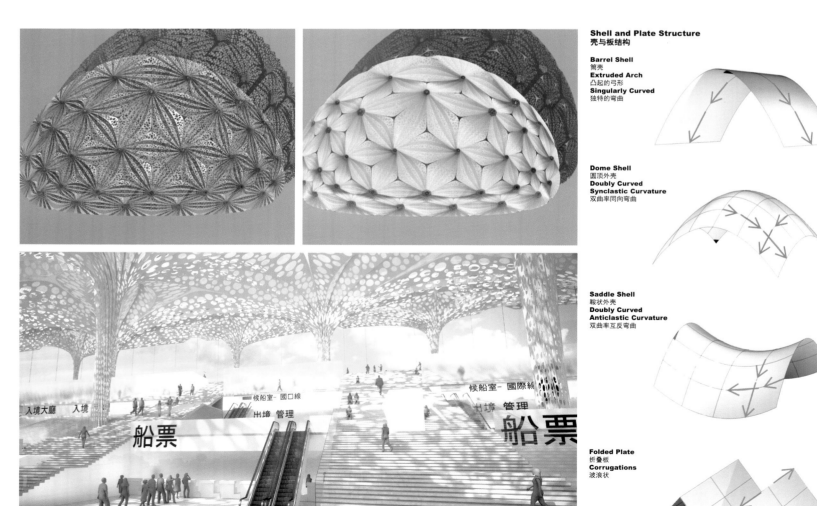

Shell and Plate Structure
壳与板结构

Barrel Shell
筒壳
Extruded Arch
凸起的弓形
Singularly Curved
独特的弯曲

Dome Shell
圆顶外壳
Doubly Curved
Synclastic Curvature
双曲率同向弯曲

Saddle Shell
鞍状外壳
Doubly Curved
Anticlastic Curvature
双曲率互反弯曲

Folded Plate
折叠板
Corrugations
波浪状

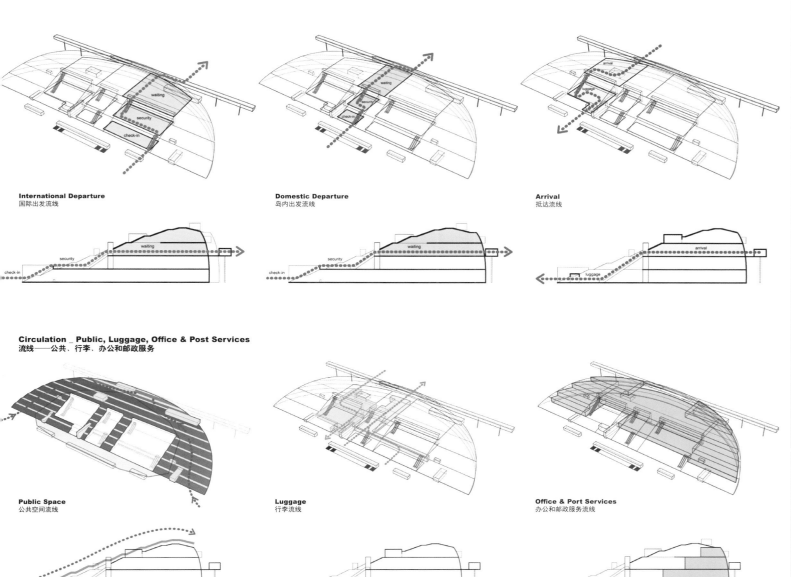

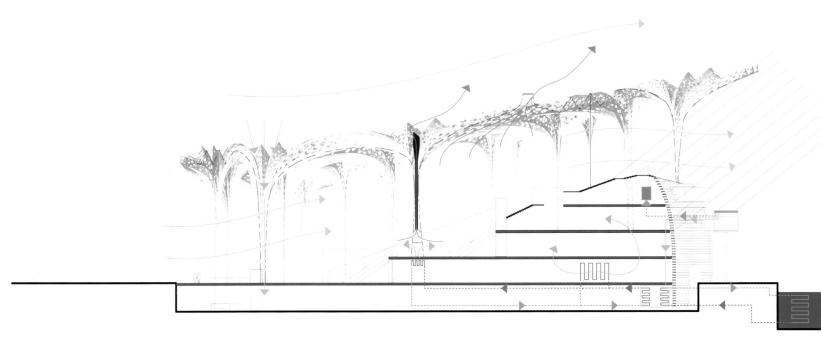

Climatic Zone 气候区	**Water Source Heat Pump Increases Air Conditioning Efficiency** 水源热泵提升空调效率		**Sunlight** 日光
Conditioned Zone 空气调节区	**Sea is Heat Sink for Water Source Heat Pump** 海洋是水源热泵的散热器		**Evacuated Solar Tubes for Domestic Hot Water** 家用热水排空太阳能管
Shaded Zone 荫蔽区	**Cooled Air for Conditioned Zone and Climatic Zone Habitation Level** 为空气调节区和气候区居住水平面提供的冷空气		**Photovoltaic Panels Supplement Grid Electricity** 光电板补充电网供电
Cross Ventilation through Shell Lace Columns 穿过缀合柱的交叉通风	**Coolant Flow / Returns** 冷却液流／返回		**Rain Collection for Grey Water Recycling** 收集雨水以对灰水回收再利用
Hot Stale Air Extracted through Natural Stack Effect from Higher Levels and Via Thermal Chimneys in Shell Lace Columns from Habitation Level 依据较高水平面自然"烟囱效应"以及通过居住层缀合柱之间的热风筒排出的浑浊热空气			

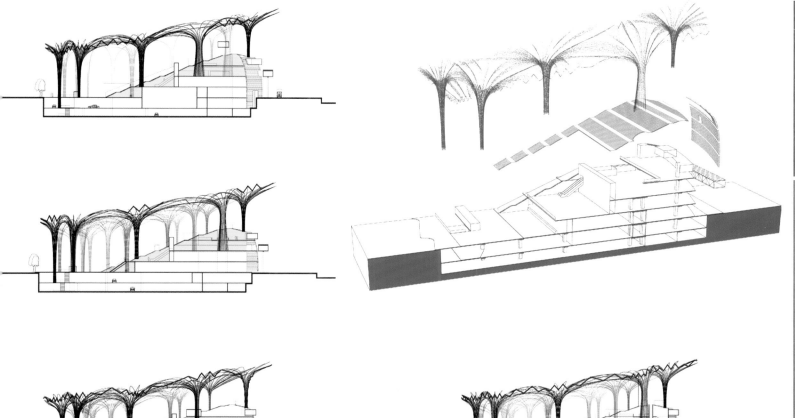

黎巴嫩贝鲁特码头 & 城镇港口
Beirut Marina & Town Quay

设计单位：斯蒂文·霍尔建筑师事务所 L.E.FT	Designed by: Steven Holl Architects; L.E.FT
合作单位：Nabil Gholam Architecture	Collaboration: Nabil Gholam Architecture
开发商：Solidere	Client: Solidere
项目地址：黎巴嫩贝鲁特	Location: Beirut, Lebanon
建筑面积：20 439 ㎡	Building Area: 20,439 m²
设计团队：Steven Holl Tim Bade Masao Akiyoshi Edward Lalonde JongSeo Lee Brett Snyder Makram El-Kadi Mohamad Ziad Jamaleddine	Design Team: Steven Holl, Tim Bade, Masao Akiyoshi, Edward Lalonde, JongSeo Lee, Brett Snyder, Makram El-Kadi, Mohamad Ziad Jamaleddine

Sea Waves, Pool Fountain, "Urban Beach"

层叠海浪 泳池瀑布 "城市沙滩"

设计构思

在水陆接壤的地方，常年海浪的拍打与侵蚀，使这些海滩的地貌在水平方向上形成了与纵向肌理截然不同的、层次鲜明的横向纹理。也正是层叠的海浪以及这种由海浪冲刷而成的地貌激发了设计师构建这个分层的水平空间的设计概念。

设计特色

黎巴嫩贝鲁特码头&城镇港口综合发展项目位于贝鲁特城镇中心，设想中的新建体量将既有的结构沿着贝鲁特海滨大道一直伸至海边，创造了一个可以观看的滨海公共场所——"城市沙滩"，也形成了一系列叠层的平台。

平台的韵律感通过整体结构上5个不同角度的弧线而形成，5条弧线的方向对应了5个泳池，故在高度的细微差别下，这些平台层和泳池也稍许倾斜，形成了因重力而产生的小瀑布，并巧妙地连接了不同的泳池层面。

为了呼应海平线，台阶选用当地的石料进行装饰，上升的平台则采用简单的几何造型，相比之下，下沿的餐厅在色彩的衬托下，显得更具活力。另外，构建在屋顶的公共观景平台也是设计的特色。

Design Concept

As the planar lapping waves of the sea inspire striated spaces in horizontal layers distinct from vertical objects, the horizontal and the planar become a geometric force shaping the new Harbor spaces.

Design Feature

Beirut Marina & Town Quay is located in Beirut town center. The envisaged new volume extends the existing structure from Beirut's Corniche to the sea to create an "Urban Beach" of public spaces overlooking the Marina and a series of stacked platforms.

The syncopated rhythm of platforms is achieved by constructing the overall curve of the Corniche in 5 angles related to the 5 reflection pools. Due to the variations in height along the Corniche, the platform levels and pools vary slightly in height allowing quiet, gravity-fed fountains to connect each pool level.

Celebrating the sea horizon, the terraces are sculpted in local stone. The simple geometry of the upper platforms is in contrast to the colorful activity of restaurants below. The building roof forms a public observation platform for the sea horizon.

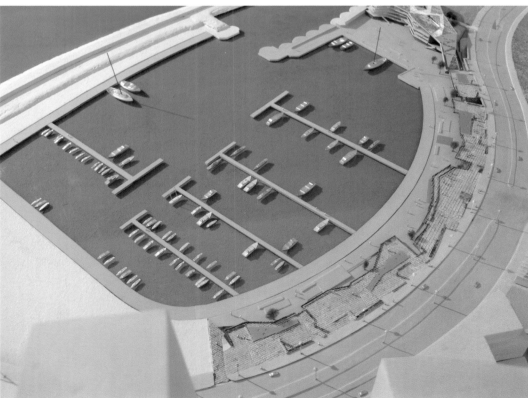

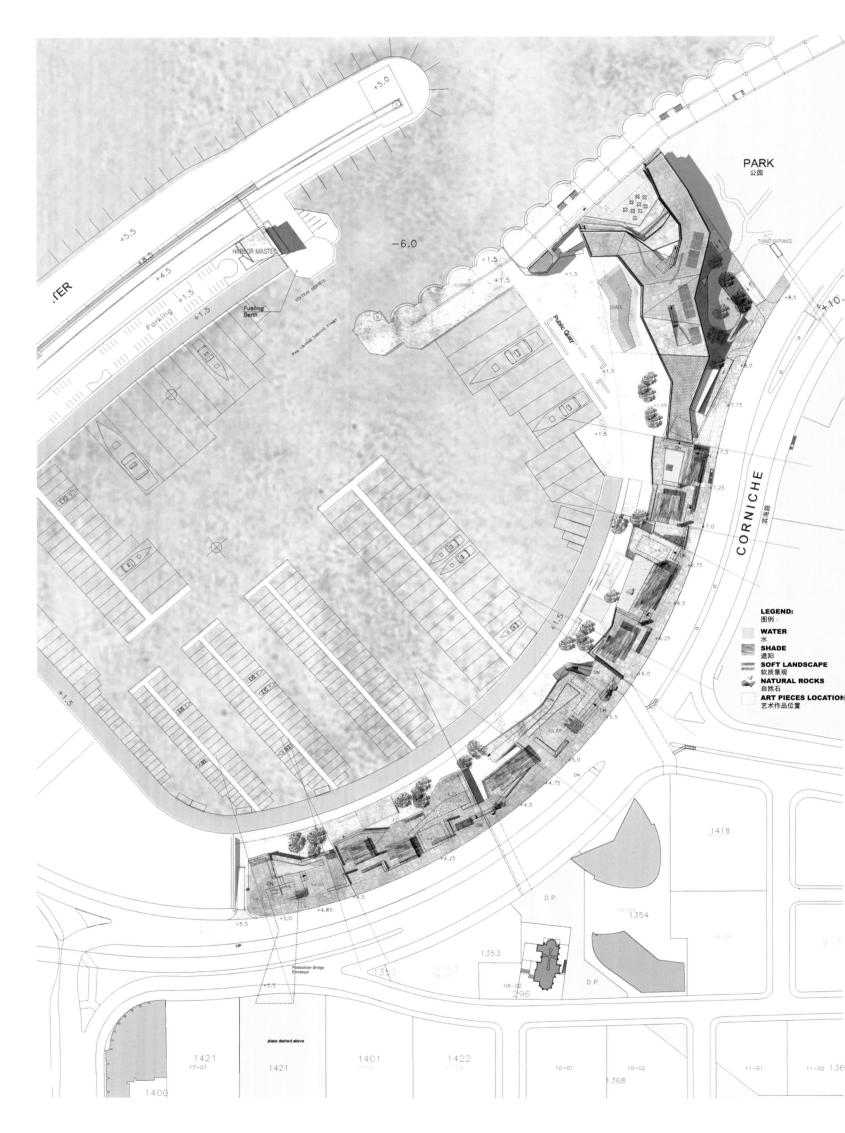

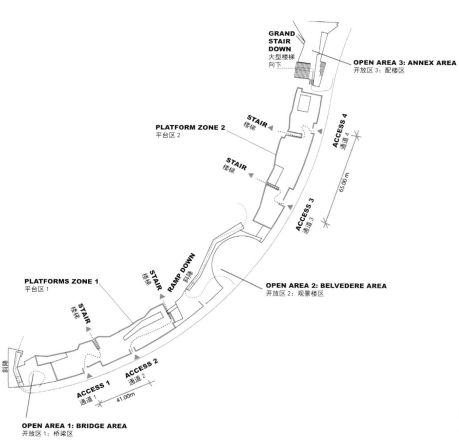

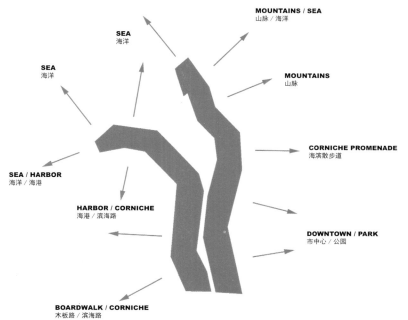

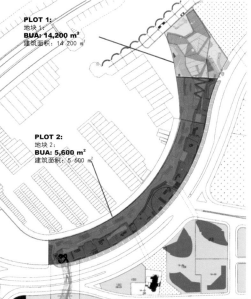
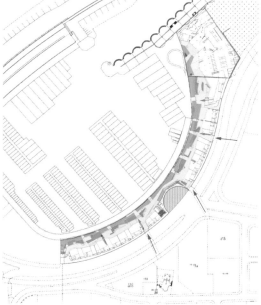
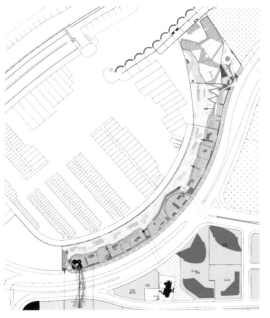

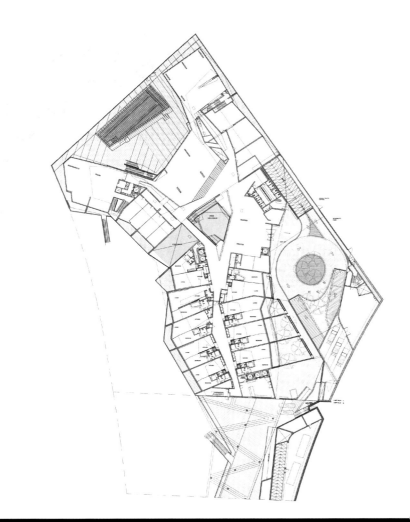

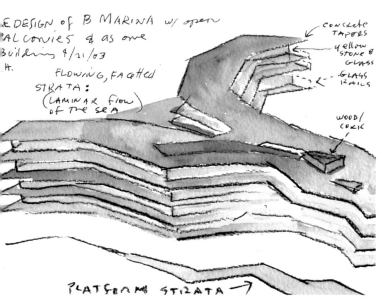
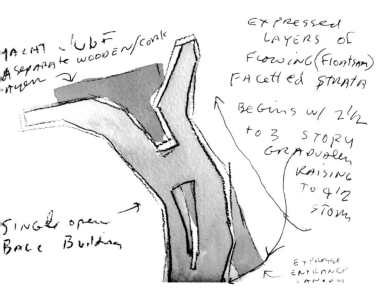
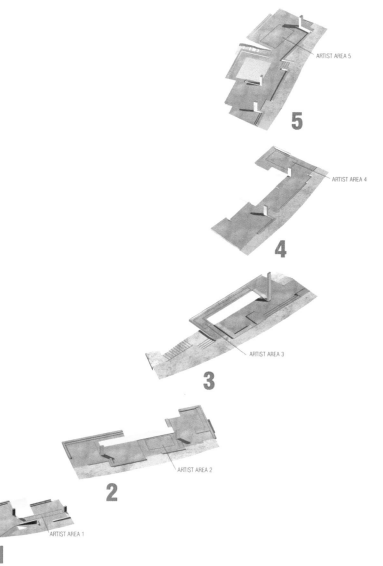

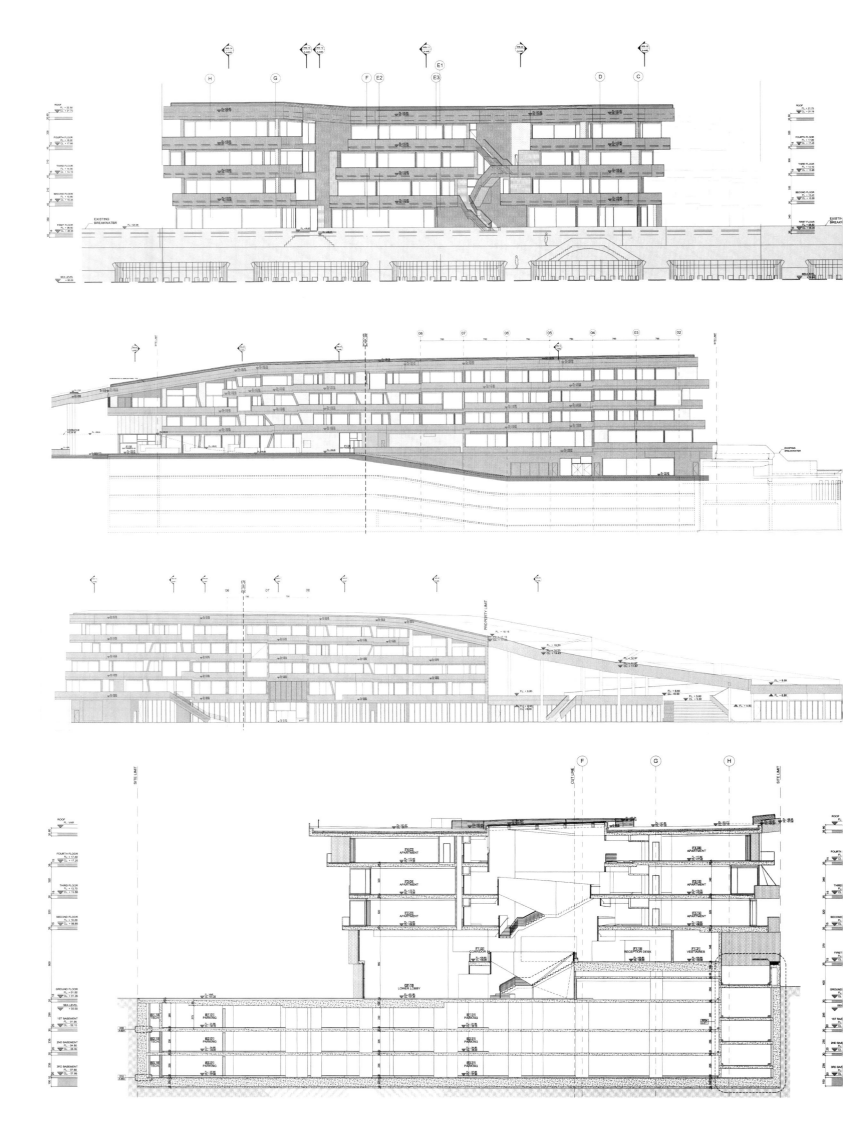

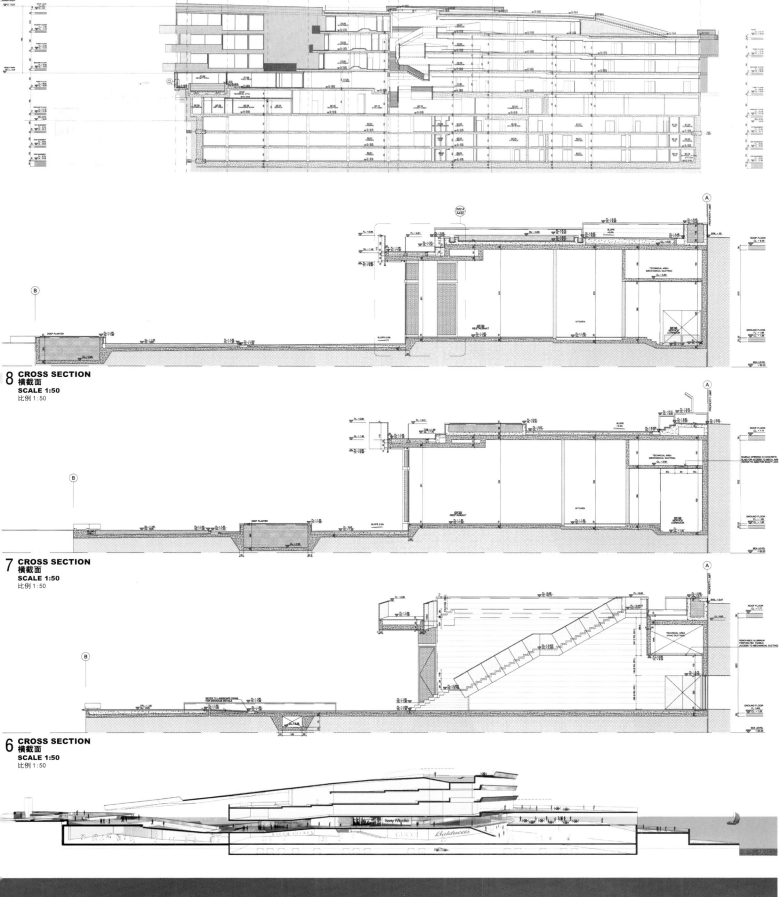

8 CROSS SECTION
横截面
SCALE 1:50
比例 1:50

7 CROSS SECTION
横截面
SCALE 1:50
比例 1:50

6 CROSS SECTION
横截面
SCALE 1:50
比例 1:50

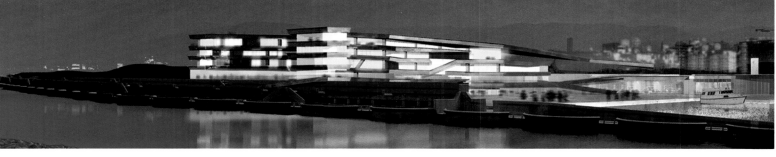

Culture & Art Building

"动物背脊"
魔法窗户
浩瀚宇宙

"Animal's Back",
Magic Window,
Vast Universe

荷兰埃曼剧院及动物园
Emmen Theater and Zoo

设计单位：Henning Larsen Architects
　　　　　Van den Berg Groep
开发单位：埃曼市
项目地址：荷兰埃曼
总建筑面积：16 000 m²
设计团队：Troels Troelsen　　Martin Stenberg
　　　　　Thomas Ponds　　Mohammad Wesam Al Asali
　　　　　Adam Ciuk　　Joakim Allerth
　　　　　Dick van de Merwe　　André Lageweg

Designed by: Henning Larsen Architects;
Van den Berg Groep
Client: City of Emmen
Location: Emmen, Netherlands
Gross Floor Area: 16,000 m²
Design Team: Troels Troelsen, Martin Stenberg,
Thomas Ponds, Mohammad Wesam Al Asali,
Adam Ciuk, Joakim Allerth,
Dick van de Merwe, André Lageweg

设计构思

这是一个融合了自然与文化的独特建筑,它既是一个大型剧院,也是当地动物园的入口空间。设计师将建筑设想为有着弧形屋顶的几何结构,弧形的屋顶让人联想到动物的背部,既呼应了建筑主题,也形成了城市独特的天际线。

另一方面,为将差异明显的城市景观与自然景观融合起来,设计将弧形屋顶与矩形框架相互渗透,这样,从城市的视角看去,建筑好似打开动物园的魔法窗户;反之,从动物园的视角看去,城市好比一个看不到边界的宇宙。

建筑设计

构成矩形框架的立面形成了不同的特征:东西立面有着大型的垂悬结构;其余立面皆为实体,其中北立面以大型窗体为特色,南立面则设有小型的开口。

剧院的空间设计极具灵活性,可举行多类型的活动。剧院的墙面可以拆除,将小舞台打开,与室外空间一起,组成一个新的户外舞台。此外,建筑的主要通道区与大厅阳台连接,形成朝向城市开放的华丽舞台。

可持续性设计

可持续性设计是埃曼新剧院开发过程中一项重要的设计规范。对日光的合理利用、几何形态与空间的灵活布局,使建筑在没有先进技术辅助的情况下,也能实现建筑节能,这种高效节能的设计方式本身也成为建筑可持续性概念的重要组成部分。

东西立面上的大型悬垂部分使广场空间免受烈日和雨水的侵袭;屋顶大量的顶灯为门厅和大厅提供了充足的日光;灵活的空间布局使不同尺寸的房间可以根据不同的需求进行调整,减少了供热和制冷所需能耗;大型的门厅和会议空间可作为自然通风的缓冲区,这些被动式设计手法,确保了良好的室内环境。

剧院的屋顶景观在减少能源需求方面起到了关键性的作用;收集到的雨水既可为整个混凝土建筑结构降温,又可用于屋顶花园的灌溉、补给运河流量及湖泊水体,或是汇入建筑前方的喷泉。

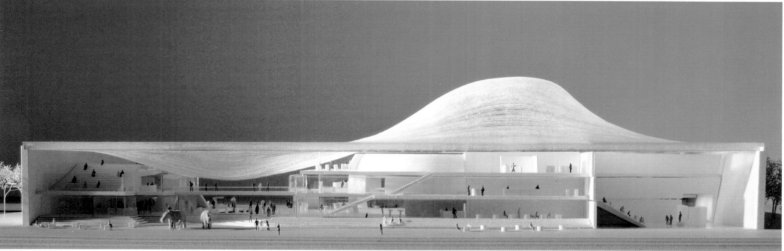
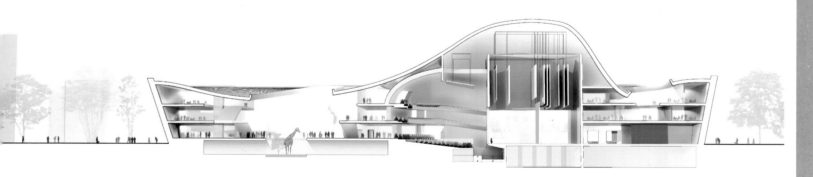
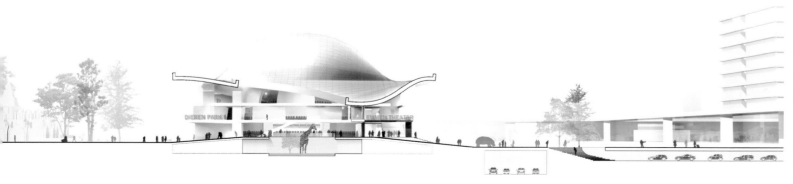

Design Concept

Emmen Theatre and Zoo is an unusual cultural building that brings together culture and nature. The building will be a large theatre and constitute the entrance to a large zoological park. It is envisaged to be a geometry with curved roof which arouses associations of an animal's back. It creates a distinctive entrance to the zoo and stands out as part of the city skyline.

On the other hand, the contrast between city and nature are united in the building design where the rectangular frame structure and the organic, curved roof merge into each other. Viewed from the city, the building opens up as a magic window towards the natural wildlife in the zoo – while the city is staged as a large urban universe viewed from the zoo.

Architectural Design

The frame structure façades have distinctive features: the large overhangs of the east and west façades; the remaining façades are solid and feature large windows to the north and small openings to the south.

Flexible space design allows the theatre to stage many different activities. The walls can be moved, and it is possible to open up the small stages and thus create new outdoor stages. In addition, the main passage area of the building situated in connection with the foyer balconies constitutes a magnificent stage opening up to city life.

Sustainable Design

Sustainability has been an important design parameter in the development of the new theatre in Emmen. The integration of daylight, geometry and flexibility in the building reduces the energy consumption without use of technology. The design is energy-efficient in itself and forms an important part of the sustainability concept of the building.

The large overhangs of the east and west façades protect the square from the sun and rain; plenty of overhead lights in the roof provide the various foyers and halls with ample daylight; the flexible design allows for the sizes of the various rooms to be adjusted as needed and reduces the energy consumption for heating or cooling; the large foyer and meeting space serves as a buffer for natural ventilation. These passive design techniques ensure favorable interior environment.

The roof landscape of the theatre plays an important role in terms of reducing energy costs. The collection of rainwater cools down the entire structure through the concrete. At the same time, the roof collects water for watering the roof garden and for filling up the small canals and lakes, which – as small water falls – fall from the roof into the fountain in front of the building.

埃塞俄比亚亚的斯亚贝巴国家体育馆和体育村

National Stadium and Sports Village

设计单位：LAVA
合作单位：DESIGNSPORT
　　　　　JDAW Architects
开发单位：埃塞俄比亚国际运动委员会
项目地址：埃塞俄比亚亚的斯亚贝巴
项目面积：600 000 ㎡

Designed by: LAVA
Collaboration: DESIGNSPORT; JDAW Architects
Client: The Federal Sports Commission of Ethiopia
Location: Addis Ababa, Ethiopia
Area: 600,000 ㎡

篮子
陨石坑
咖啡豆
远古竞技场

Basket,
Meteor Crater,
Coffee Bean,
Arena

项目概况

项目位于埃塞俄比亚亚的斯亚贝巴，是为举办全球性的足球运动比赛而设计的场馆，包括可容纳60 000人的主场馆、水上运动场地以及竞技场。

设计特色

在设计过程中，设计师尽量回归到建筑设计的本源，采用最简单、最自然的材料来建筑一个亲切的体育馆。设计的灵感来源于对远古竞技场的回溯，这个类似于陨石坑，又像咖啡豆的建筑是对现有地形的重塑，既创造了高效的空间，优化了环保性能，又最大限度地降低了建设成本，并集成了现有的景观设施。

在建筑风格上，该建筑参照了埃塞俄比亚的特色建筑风格，隐隐流露出当地知名教堂和民居的韵味，实现了建筑与地区文化的完美融合。

在建筑形态上，受当地传统食物篮子的启发，整个建筑的外表面由斑斓的镶板组合而成，极具视觉冲击力，而观众看台区则是由再生土填充而成。建筑的屋顶覆盖着一层智能膜，好似飘浮在建筑之上的白云。

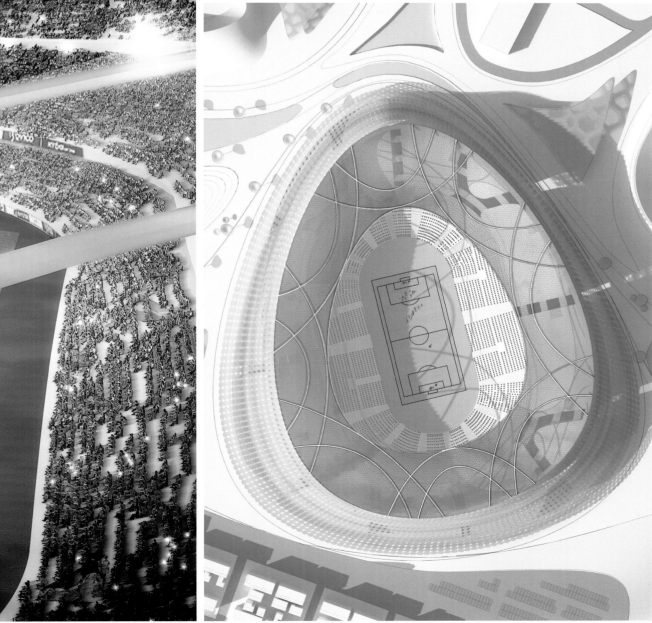

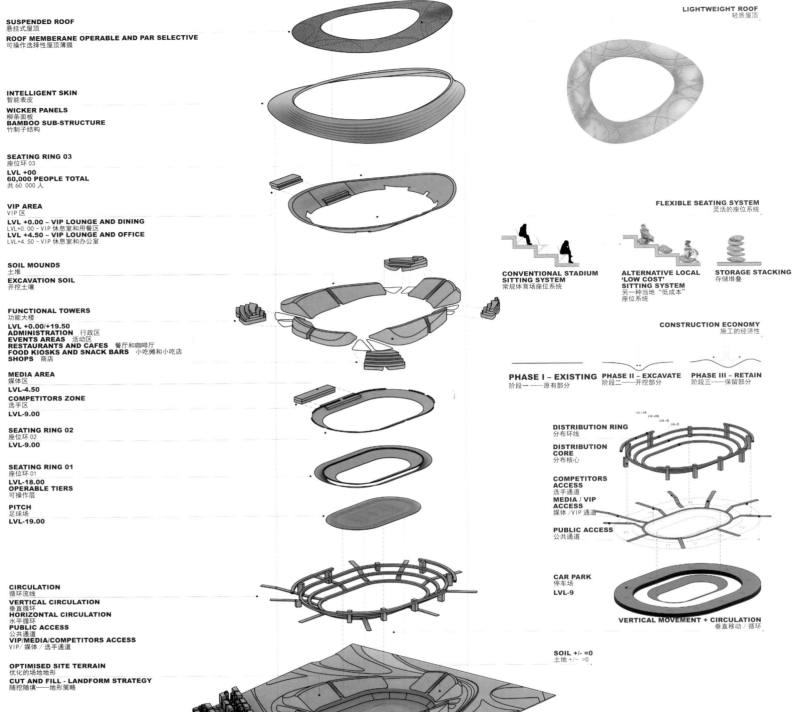

SUSPENDED ROOF
悬挂式屋顶
ROOF MEMBERANE OPERABLE AND PAR SELECTIVE
可操作选择性屋顶薄膜

INTELLIGENT SKIN
智能表皮
WICKER PANELS
柳条面板
BAMBOO SUB-STRUCTURE
竹制子结构

SEATING RING 03
座位环 03
LVL +00
60,000 PEOPLE TOTAL
共 60 000 人

VIP AREA
VIP 区
LVL +0.00 – VIP LOUNGE AND DINING
LVL+0.00 - VIP 休息室和用餐区
LVL +4.50 – VIP LOUNGE AND OFFICE
LVL+4.50 - VIP 休息室和办公室

SOIL MOUNDS
土堆
EXCAVATION SOIL
开挖土壤

FUNCTIONAL TOWERS
功能大楼
LVL +0.00/+19.50
ADMINISTRATION 行政区
EVENTS AREAS 活动区
RESTAURANTS AND CAFES 餐厅和咖啡厅
FOOD KIOSKS AND SNACK BARS 小吃摊和小吃店
SHOPS 商店

MEDIA AREA
媒体区
LVL-4.50
COMPETITORS ZONE
选手区
LVL-9.00

SEATING RING 02
座位环 02
LVL-9.00

SEATING RING 01
座位环 01
LVL-18.00
OPERABLE TIERS
可操作层
PITCH
足球场
LVL-19.00

CIRCULATION
循环流线
VERTICAL CIRCULATION
垂直循环
HORIZONTAL CIRCULATION
水平循环
PUBLIC ACCESS
公共通道
VIP/MEDIA/COMPETITORS ACCESS
VIP/ 媒体 / 选手通道

OPTIMISED SITE TERRAIN
优化的场地地形
CUT AND FILL - LANDFORM STRATEGY
随挖随填——地形策略

LIGHTWEIGHT ROOF
轻质屋顶

FLEXIBLE SEATING SYSTEM
灵活的座位系统

CONVENTIONAL STADIUM SITTING SYSTEM
常规体育场座位系统

ALTERNATIVE LOCAL 'LOW COST' SITTING SYSTEM
另一种当地"低成本"座位系统

STORAGE STACKING
存储堆叠

CONSTRUCTION ECONOMY
施工的经济性

PHASE I – EXISTING
阶段一——原有部分
PHASE II – EXCAVATE
阶段二——开挖部分
PHASE III – RETAIN
阶段三——保留部分

DISTRIBUTION RING
分布环线
DISTRIBUTION CORE
分布核心
COMPETITORS ACCESS
选手通道
MEDIA / VIP ACCESS
媒体 /VIP 通道
PUBLIC ACCESS
公共通道

CAR PARK
停车场
LVL-9

VERTICAL MOVEMENT + CIRCULATION
垂直移动 / 循环

SOIL +/- =0
土地 +/- =0

Profile

The project located in Addis Ababa, Ethiopia is a new national stadium for international football match. It includes a 60,000 seat stadium, water sports venue and arena.

Design Feature

Designers have gone back to the very origin of stadium design, taking the simplest and most natural materials to build a friendly stadium. The design inspiration originates from ancient arena.

The form of the stadium structure also recalls coffee beans and meteor crater. This man-made crater is a clever remodeling of the existing terrain and generates efficient spaces, optimizes environmental performance, minimizes construction costs and integrates facilities within the existing landscape.

The design references Ethiopia's world-famous excavated architecture – centuries-old rock churches, dwellings and cisterns, achieving a perfect combination of the building and local culture.

The Massob, an Ethiopian communal serving basket made from woven grass, inspired the façade material that wraps the stadium. The grandstands around a sunken arena are formed from excavated material. The roof of the stadium, an intelligent membrane, appears like a cloud floating over the building.

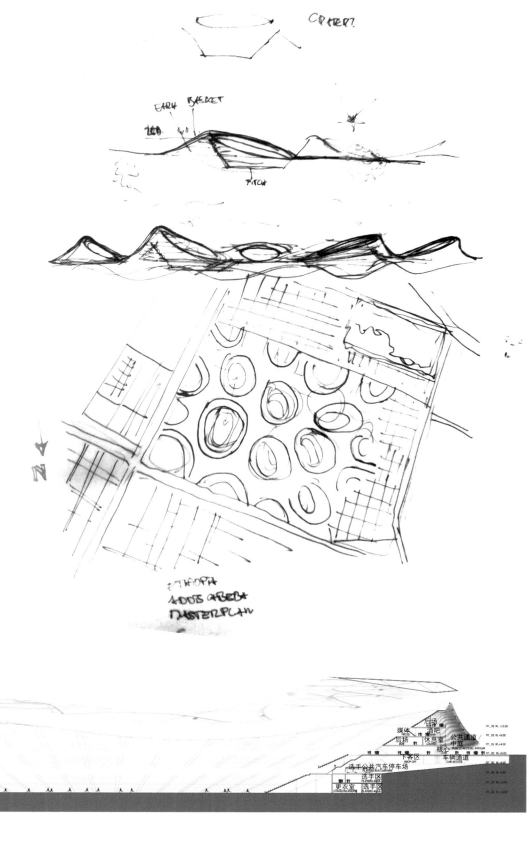

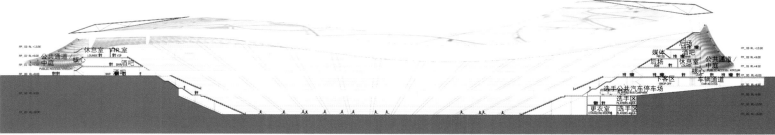

SECTION A-A'
剖面 A-A'

SECTION B-B'
剖面 B-B'

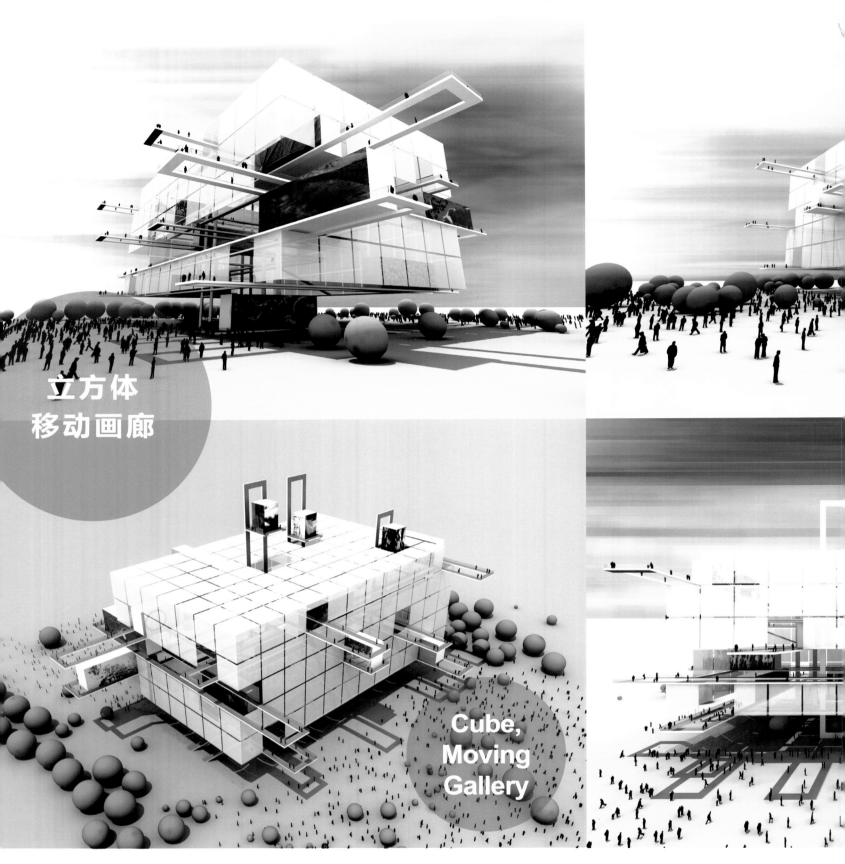

台湾新北艺术博物馆

New Taipei City Museum of Art

设计单位：Process-based Architecture Studio
开发单位：新北市政府
项目地址：中国台湾省新北市
设计团队：Jafar Bazzaz
　　　　　Arash Pouresmaeil
　　　　　Kamal Youssefpoor

Designed by: Process-based Architecture Studio
Client: New Taipei City Government
Location: New Taipei City, Taiwan, China
Design Team: Jafar Bazzaz,
Arash Pouresmaeil, Kamal Youssefpoor

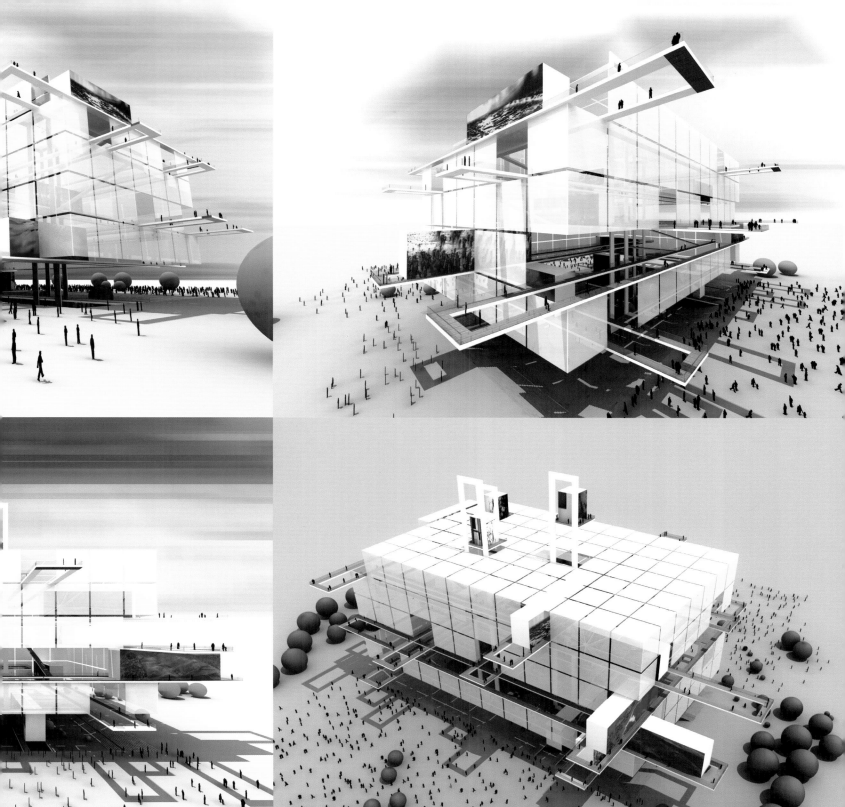

项目概况

新北艺术博物馆设计方案在城市广场上设想了一个大型的立方体,这一立方体将融艺术、生活、娱乐于一体,表现艺术与当代台湾文化生活之间的本质联系,树立强烈的文化认同感。

设计特色

这个悬浮立方体的内部是一个集成的空间,没有像一般的建筑一样分成特定的楼层。其内设有多个可移动的画廊,可水平移动或是垂直移动,是展示艺术品的上好空间。另外,这些可移动的画廊可穿越建筑的 4 个主要立面,甚至是被称为第 5 立面的屋顶,并沿着立面的轨迹融入城市空间。

移动画廊在建筑的内部和外部营造了一个动态的空间,这一空间还会随着时间的推移而不断变化。博物馆的组织结构因画廊的运动而定义,因此,当人们穿过博物馆时,可以观赏到不同的立面,并在不同的时间段内获得不同的空间体验,给人如观看影片般的感触。

Profile

The proposal for New Taipei City Museum of Art includes a huge cube on an urban plaza. The cube integrates art with life and entertainment to show a connection between art and contemporary cultural life in Taiwan and achieve strong cultural identity.

Design Feature

The interior space of this suspended cube is integrated and has not been divided to specified floors as the usual buildings. It consists of several moving galleries (move horizontally or vertically) where exterior walls of them are also good places for displaying the art works. The moving galleries of museum are able to cross the building's exterior wall in four main façades and the fifth one (roof), on their path and to enter the urban space.

Moving galleries create a dynamic space inside and outside the museum that changes throughout time and its organization depends on the galleries' motions. Therefore, people passing the museum building, will witness different façades and visitors of the museum will experience different spaces inside, in various times. This spatial experience has unique cinematic and dramatic character.

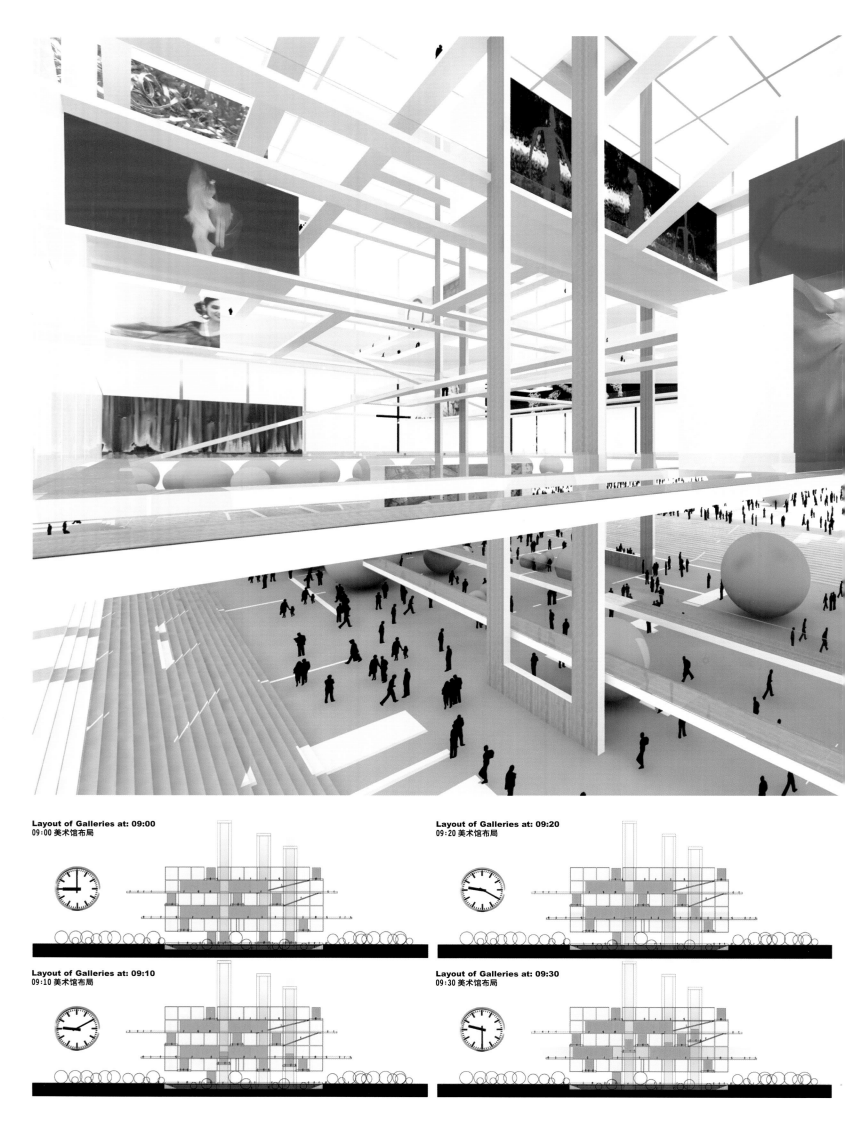

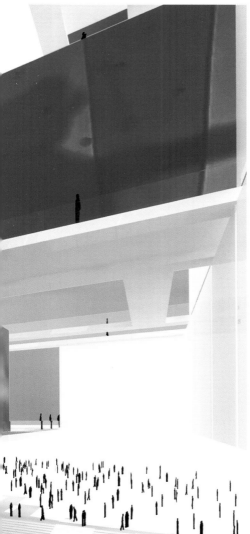

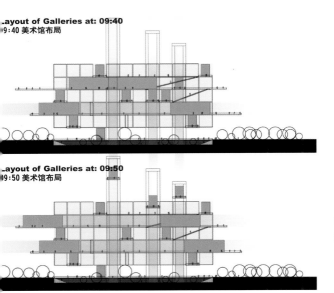

Layout of Galleries at: 09:40
09:40 美术馆布局

Layout of Galleries at: 09:50
09:50 美术馆布局

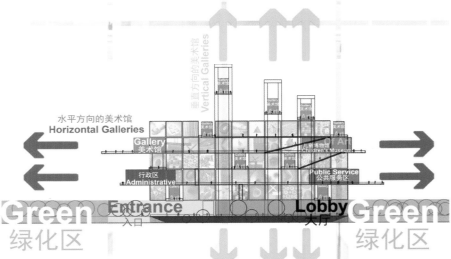

垂直方向的美术馆 Vertical Galleries
水平方向的美术馆 Horizontal Galleries
Gallery 美术馆
Children's Museum 儿童博物馆
行政区 Administrative
Public Service 公共服务区
Green 绿化区
Entrance 入口
Lobby 大厅
Green 绿化区

美国纽约韩国文化中心
Korean Cultural Center in New York

设计单位：OBRA Architects
项目地址：美国纽约

Designed by: OBRA Architects
Location: New York, USA

设计构思

提到美国，脑中不由地冒出这些词：开放、兼容，这一点在文化上表现得更为突出，也正因为美国高度的包容性，使得多种文化可以在同一空间内共存，那么，在纽约构建一个展示韩国文化和形象的文化中心，也就成为一种可能。

纵然韩国文化中心也像其他建筑一样，位于纽约第32号街，但设计师给它的定义却更为深远——韩国送给纽约市的文化礼物，在这个空间里，韩国人民、美国人民以及其他国家的人们可以分享韩国在文化领域的成就，并共同展望它在未来的演变，因此，设计师将建筑的主题定为：过去和未来。

考虑到这一项目的独特性，设计师没有采用美国建筑、韩国建筑或其他国家的建筑形式来诠释这一项目，而是通过文化的融合，形成一种新的风格，就像铜和锡融合可以产生青铜一样，以融合两国文化特色的设计手法设想了一个独特的原创空间，从而赋予这一建筑独特的"声音"和"符号"。

建筑设计

这一建筑是由两个体量组成的整体，访客可以从下至上沿着空间和活动的既定序列进行探索，或者反过来从上往下参观，从而获得另一种体验。这样多元的诠释方式，使建筑就像一本展示在人们眼前、可以以任何顺序来阅读的书一样，给予访客更多的自由和选择。

建筑的玻璃外立面设计成"薄云"的几何形态，这些"薄云"由酸蚀玻璃和钢构件组成，将为韩国文化中心的每一个空间带来独特的特征，同时也在该市的建筑群中，赋予整个建筑无与匹敌的显著特征。在建筑体量的复合模式中，这些"薄云"暂时被搁置在建筑墙体中，成为新旧韩国文化的参照。同时，这些玻璃立面也让人们联想到现代韩国文化的多样性。

"书"
"薄云"
文化融合

"Book",
"Light Clouds",
Cultural Integration

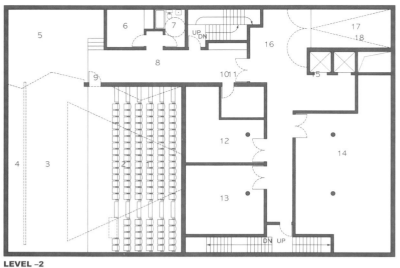

LEVEL –2
–2层
–6.70m (–22'0")

1. **AUDITORIUM**
   1. 礼堂
2. **HANDICAP SEATING**
   2. 无障碍座位
3. **AUDITORIUM STAGE**
   3. 礼堂舞台
4. **CROSSOVER**
   4. 天桥
5. **BACK STAGE**
   5. 后台
6. **WAITING ROOM**
   6. 休息室
7. **UNISEx BATHROOM**
   7. 卫生间
8. **STAGING AREA**
   8. 表演区
9. **HANDICAP DOOR / EMERGENCY Exit**
   9. 无障碍门 / 紧急出口
10. **BENCH**
    10. 长凳
11. **STAGE OFFICE**
    11. 舞台办公室
12. **BUILDING STORAGE**
    12. 建筑仓库
13. **ExHIBITION STORAGE**
    13. 展览仓库
14. **MECHANICAL ROOM**
    14. 机械房
15. **ELEVATOR LOBBY**
    15. 电梯门厅
16. **LOADING AREA**
    16. 卸货区
17. **VEHICLE ELEVATOR**
    17. 车辆电梯
18. **MECHANICAL SHAFT**
    18. 机械轴

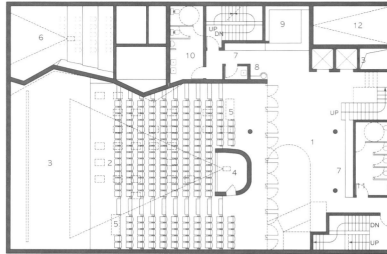

LEVEL –1
–1层
–3.35m (–11'0")

1. **AUDITORIUM LOBBY**
   1. 礼堂大厅
2. **AUDITORIUM**
   2. 礼堂
3. **STAGE BELOW**
   3. 下方舞台
4. **AV ROOM**
   4. 影音室
5. **HANDICAP SEATING**
   5. 无障碍座位
6. **LECTURE HALL**
   6. 演讲厅
7. **BENCH**
   7. 长凳
8. **DRINKING FOUNTAIN**
   8. 自动饮水器
9. **CONCESSION STAND**
   9. 小卖部
10. **MENS RESTROOM**
    10. 男卫生间
11. **WOMENS RESTROOM**
    11. 女卫生间
12. **VEHICLE ELEVATOR**
    12. 车辆电梯

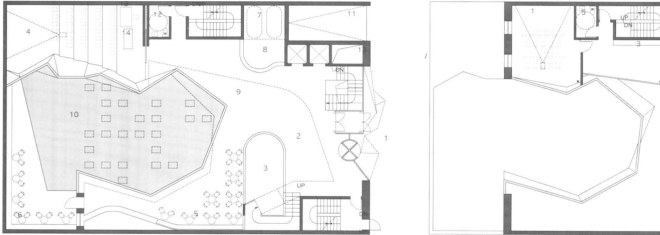

GROUND LEVEL
底层
+0.00m (+0'0")

1. **ENTRY**
   1. 入口
2. **STREET LOBBY**
   2. 街道大厅
3. **OPEN TO AUDITORIUM LOBBY BELOW**
   3. 向下方礼堂大厅开敞
4. **LECTURE HALL**
   4. 演讲厅
5. **CAFÉ**
   5. 咖啡馆
6. **CAFE TERRACE**
   6. 咖啡露台
7. **COAT CHECK CONVEYOR**
   7. 衣帽间运输设备
8. **INFORMATION**
   8. 信息
9. **SHOW WINDOW ExHIBITION SPACE**
   9. 橱窗展览空间
10. **REFLECTING POOL**
    10. 倒影池
11. **VEHICLE ELEVATOR**
    11. 车辆电梯
12. **UNISEx RESTROOM**
    12. 卫生间
13. **MECHANICAL SHAFT**
    13. 机械轴
14. **SOUND CONTROL**
    14. 声音控制

MEZZANINE LEVEL
夹层
+2.90m (+9'6")

1. **LECTURE ROOM**
   1. 演讲室
2. **OPEN TO STREET LOBBY BELOW**
   2. 向下方街道大厅开敞
3. **BENCH**
   3. 长凳
4. **DRINKING FOUNTAIN**
   4. 自动饮水器
5. **UNISEX RESTROOM**
   5. 卫生间
6. **MECHANICAL SHAFT**
   6. 机械轴
7. **HANDICAP SEATING**
   7. 无障碍座位

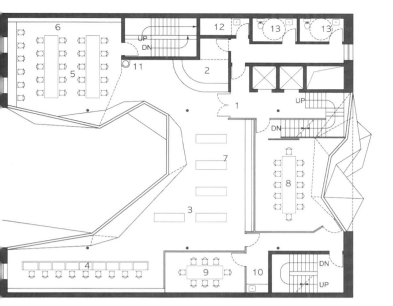

LEVEL +2
+6.40m (+21'0")

1. ELEVATOR LOBBY 电梯门厅
2. LIBRARY REFERENCE DESK 图书馆参考咨询台
3. LIBRARY STACKS 图书馆书库
4. DIGITAL BROWSING AREA 电子阅览区
5. READING ROOM 阅览室
6. 12-PERSON MEETING ROOM / GROUP STUDY ROOM 12人会议室 / 小组讨论室
7. 8-PERSON MEETING ROOM / GROUP STUDY ROOM 8人会议室 / 小组讨论室
8. LIBRARIAN STORAGE 图书馆仓库
9. SKYLIGHT ABOVE 上方天窗
10. DRINKING FOUNTAIN 自动饮水器
11. SERVICE SINK 服务水槽
12. UNISEX RESTROOM 卫生间
13. MECHANICAL SHAFT 机械轴

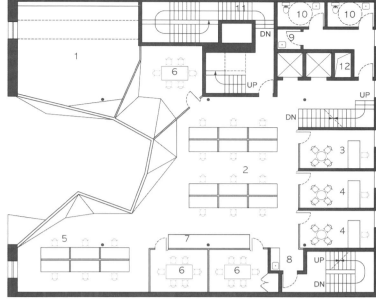

LEVEL +3
+10.40m (+34'0")

1. LIBRARY MAIN READING ROOM BELOW 下方图书馆主阅览室
2. KOREA CULTURAL CENTER OFFICES 韩国文化中心办公室
3. KOREA CULTURAL CENTER DIRECTORS OFFICE 韩国文化中心主管办公室
4. DIRECTORS OFFICE (TO BE DECIDED) 主管办公室（待定）
5. OPEN OFFICE SPACE (TO BE DECIDED) 开放型办公空间（待定）
6. 6-PERSON CONFERENCE ROOM 6人会议室
7. FILE STORAGE 文件仓库
8. TEA KITCHENETTE 茶品小厨房
9. SERVICE SINK 服务水槽
10. UNISEX RESTROOM 卫生间
11. SKYLIGHT OPENING TO LIBRARY BELOW 向下方图书馆打开的天窗
12. MECHANICAL SHAFT 机械轴

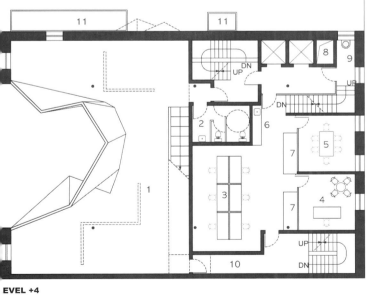

LEVEL +4
+13.10m (+43'0")

1. EXHIBITION SPACE 展览空间
2. WOMENS RESTROOM 女卫生间
3. 12-PERSON CONFERENCE ROOM 12人会议室
4. TOURISM DIRECTORS OFFICE 旅游主管办公室
5. SIX PERSON CONFERENCE ROOM 6人会议室
6. TEA KITCHENETTE 茶品小厨房
7. FILE STORAGE 文件仓库
8. MECHANICAL SHAFT 机械轴
9. DRINKING FOUNTAIN 自动饮水器
10. EMERGENCY EGRESS 紧急出口
11. SKYLIGHT TO BELOW 天窗

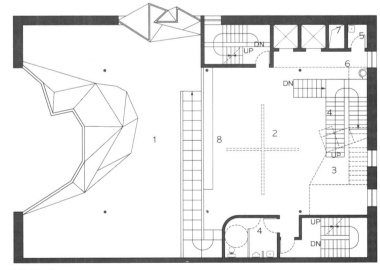

LEVEL +5
+15.80m (+52'0")

1. EXHIBITION SPACE BELOW 下方展览空间
2. EXHIBITION SPACE 展览空间
3. SKYLIGHT ABOVE 上方天窗
4. MENS RESTROOM 男卫生间
5. SERVICE SINK 服务水槽
6. DRINKING FOUNTAIN 自动饮水器
7. MECHANICAL SHAFT 机械轴
8. BENCH 长凳

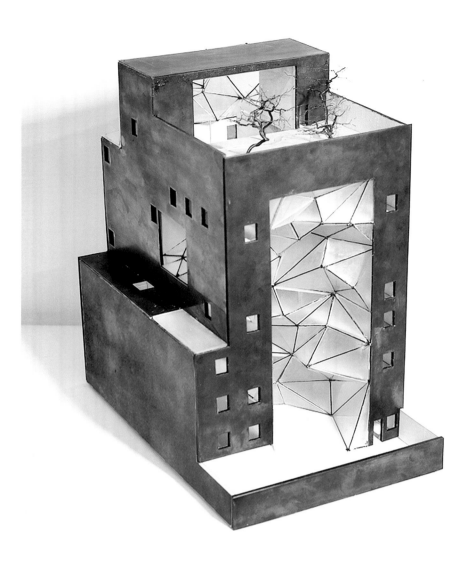

Design Concept

People may easily relate America to "open" and "tolerant", especially to the American culture which makes it possible to integrate multiple cultures in the same space. Therefore, it's not difficult to imagine a cultural center in New York showing Korean culture and image.

Although it will physically occupy its place amongst the other buildings on 32nd Street, its content will be quite something else – it's a cultural gift to New York City. It is a space where Koreans, Americans and everyone else can share together the contributions of Korea to world culture and human development so far, and also witness the unfolding of its future promise. The very substance of the Korean Cultural Center, the emblem of Korea in New York City will be this bipolar theme of past and future.

The project then, to be successful, must envision a building like no other, either in New York, Korea or anywhere else for that matter, and aim to create original spaces of new styles via integration of cultures like that wisely conceived alloy of copper and tin come together to create something that did not exist before (bronze). The project is endowed with unique "sound" and "symbol".

Architectural Design

The building is composed of two volumes. Visitors are allowed to explore it from the bottom-up with a determined sequence of spaces and activities or travel it "in reverse" from the top down for a completely different experience. Like a book that can be read forwards or backwards or in any other order, the building opens itself up to "multiple interpretations" of space in this way, and aspires to "democratically" maximize the visitor's freedom.

The glazed façades of the building have been designed as geometricized Light Clouds. They are constructed out of acid-etched glass structured with steel elements. These Clouds of Light will give unique character to each of the spaces of the Korean Cultural Center as they also endow the entire building with characteristic and memorable way of being equal to no other building in the city. In the complex patterning of their volumes, the Light Clouds, as if momentarily delayed in the walls of the building, can be seen as making reference to both old and new aspects of Korean culture. Also, these façades can suggest the "multi-faceted" diversity of contemporary Korean culture.

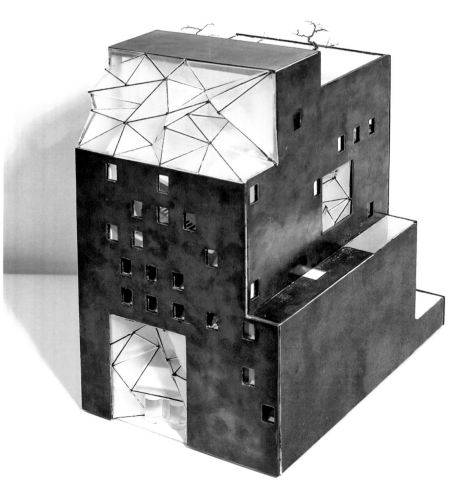

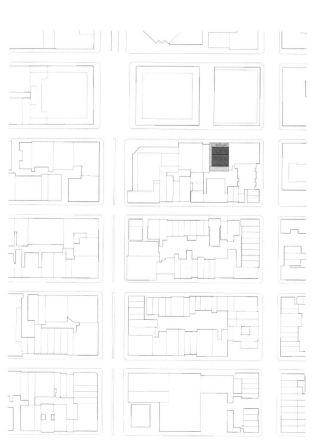

ALTHOUGH MEMORABLE AND UNIQUE, THE APPEARANCE OF THE KCC WILL HARMONIZE WITH THE CHARACTERISTIC NEW YORK CONTExT OF 32ND STREET.
韩国文化中心在拥有独特外观的同时，还将与第 32 街特色的纽约环境相协调。

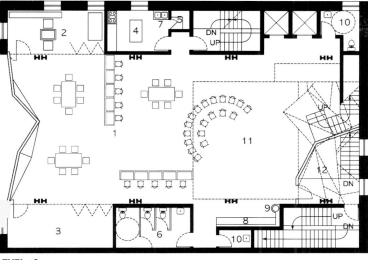
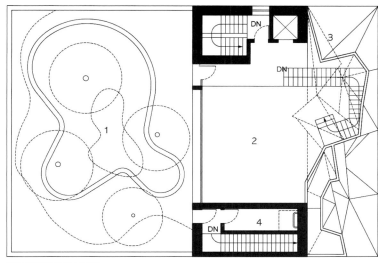

LEVEL +6
+6层
21.30m (+70'0")

1. **KOREAN EXPERIENCE HALL**
   韩国体验展厅
2. **SARANG BANG (CULTURAL HALL 1)**
   莎廊坊（文化大厅1）
3. **CULTURAL PRODUCTS (CULTURAL HALL 2)**
   文化产品（文化大厅2）
4. **FULL KITCHEN (CULTURAL HALL 3)**
   大厨房（文化大厅3）
5. **KITCHEN STORAGE**
   厨房仓库
6. **WOMEN'S RESTROOM**
   女卫生间
7. **SERVICE SINK**
   服务水槽
8. **BENCH**
   长凳
9. **DRINKING FOUNTAIN**
   自动饮水器
10. **UNISEX RESTROOM**
    卫生间
11. **ExHIBITION SPACE BELOW**
    下方展览空间
12. **OPEN TO GARDEN ABOVE**
    向上方花园打开

ROOF GARDEN LEVEL
屋顶花园层
+26.80m (+88'0")

1. **KOREAN GARDEN**
   韩国花园
2. **OPEN TO KOREAN EXPERIENCE HALL BELOW**
   向下方韩国体验厅打开
3. **SKYLINE MEDITATION CHAPEL**
   天际冥想教堂
4. **STAFF ACCESS TO MECHANICAL PENTHOUSE**
   通往屋顶机房的员工通道

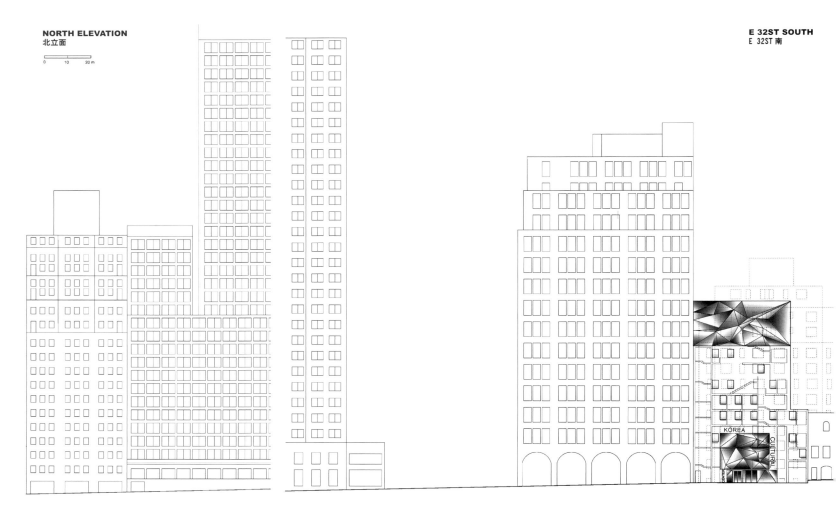

NORTH ELEVATION
北立面

E 32ST SOUTH
E 32ST 南

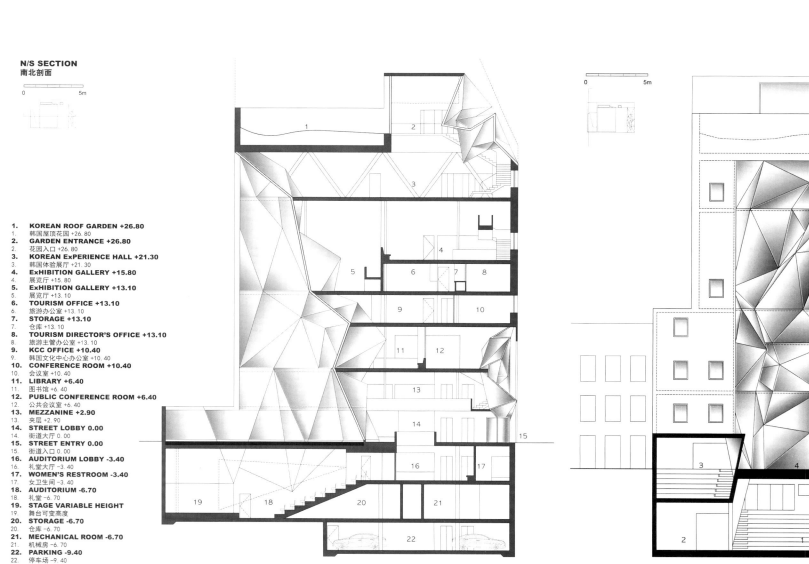

N/S SECTION
南北剖面

1. **KOREAN ROOF GARDEN +26.80**
1. 韩国屋顶花园 +26.80
2. **GARDEN ENTRANCE +26.80**
2. 花园入口 +26.80
3. **KOREAN EXPERIENCE HALL +21.30**
3. 韩国体验展厅 +21.30
4. **EXHIBITION GALLERY +15.80**
4. 展览厅 +15.80
5. **EXHIBITION GALLERY +13.10**
5. 展览厅 +13.10
6. **TOURISM OFFICE +13.10**
6. 旅游办公室 +13.10
7. **STORAGE +13.10**
7. 仓库 +13.10
8. **TOURISM DIRECTOR'S OFFICE +13.10**
8. 旅游主管办公室 +13.10
9. **KCC OFFICE +10.40**
9. 韩国文化中心办公室 +10.40
10. **CONFERENCE ROOM +10.40**
10. 会议室 +10.40
11. **LIBRARY +6.40**
11. 图书馆 +6.40
12. **PUBLIC CONFERENCE ROOM +6.40**
12. 公共会议室 +6.40
13. **MEZZANINE +2.90**
13. 夹层 +2.90
14. **STREET LOBBY 0.00**
14. 街道大厅 0.00
15. **STREET ENTRY 0.00**
15. 街道入口 0.00
16. **AUDITORIUM LOBBY -3.40**
16. 礼堂大厅 -3.40
17. **WOMEN'S RESTROOM -3.40**
17. 女卫生间 -3.40
18. **AUDITORIUM -6.70**
18. 礼堂 -6.70
19. **STAGE VARIABLE HEIGHT**
19. 舞台可变高度
20. **STORAGE -6.70**
20. 仓库 -6.70
21. **MECHANICAL ROOM -6.70**
21. 机械房 -6.70
22. **PARKING -9.40**
22. 停车场 -9.40

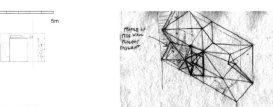

E/W SECTION
东西剖面

1. **MECHANICAL PENTHOUSE +29.70**
1. 屋顶机械房 +29.70
2. **GARDEN ENTRANCE +26.80**
2. 花园入口 +26.80
3. **EMERGENCY EGRESS**
3. 紧急出口
4. **KOREAN ExPERIENCE HALL +21.30**
4. 韩国体验展厅 +21.30
5. **SERVICE SINK**
5. 服务水槽
6. **MECHANICAL SPACE**
6. 机械空间
7. **ExHIBITION GALLERY +15.80**
7. 展览厅 +15.80
8. **MEN'S RESTROOM +15.80**
8. 男卫生间 +15.80
9. **WONEN'S RESTROOM +13.10**
9. 女卫生间 +13.10
10. **TOURISM OFFICE +13.10**
10. 旅游办公室 +13.10
11. **KCC OFFICE +10.40**
11. 韩国文化中心办公室 +10.40
12. **CONFERENCE ROOM +10.40**
12. 会议室 +10.40
13. **LIBRARY ENTRANCE LOBBY +6.40**
13. 图书馆入口大厅 +6.40
14. **CONFERENCE ROOM/LIBRARY STUDY ROOM +6.40**
14. 会议室／图书馆自习室 +6.40
15. **MEZZANINE +2.90**
15. 夹层 +2.90
16. **LECTURE HALL +2.90**
16. 演讲厅 +2.90
17. **STREET ENTRY LOBBY 0.00**
17. 街道入口大厅 0.00
18. **MECHANICAL COAT CHECK /INFORMATION DESK 0.00**
18. 机械存么／信息台 0.00
19. **AUDITORIUM LOBBY -3.40**
19. 礼堂大厅 -3.40
20. **CONCESSION STAND / TICKET BOOTH -3.40**
20. 小卖部／售票处 -3.40
21. **BENCH**
21. 长凳
22. **LOADING AREA -6.70**
22. 装载区 -6.70
23. **STORAGE AREA -6.70**
23. 仓库区 -6.70
24. **PARKING -9.40**
24. 停车场 -9.40

S ELEVATION
南立面

1. **AUDITORIUM -3.35 TO -6.70**
1. 礼堂 -3.35——-6.70
2. **BACK STAGE -6.70**
2. 后台 -6.70
3. **LECTURE HALL -2.40**
3. 演讲厅 -2.40
4. **REFLECTIVE POOL 0.00**
4. 倒影池 0.00
5. **CAFE TERRACE ROOM 0.00**
5. 咖啡露台 0.00

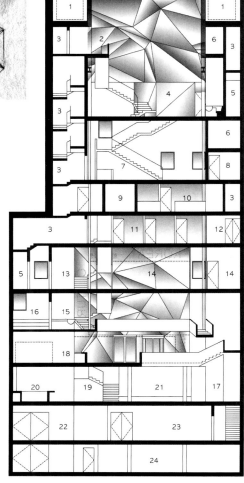

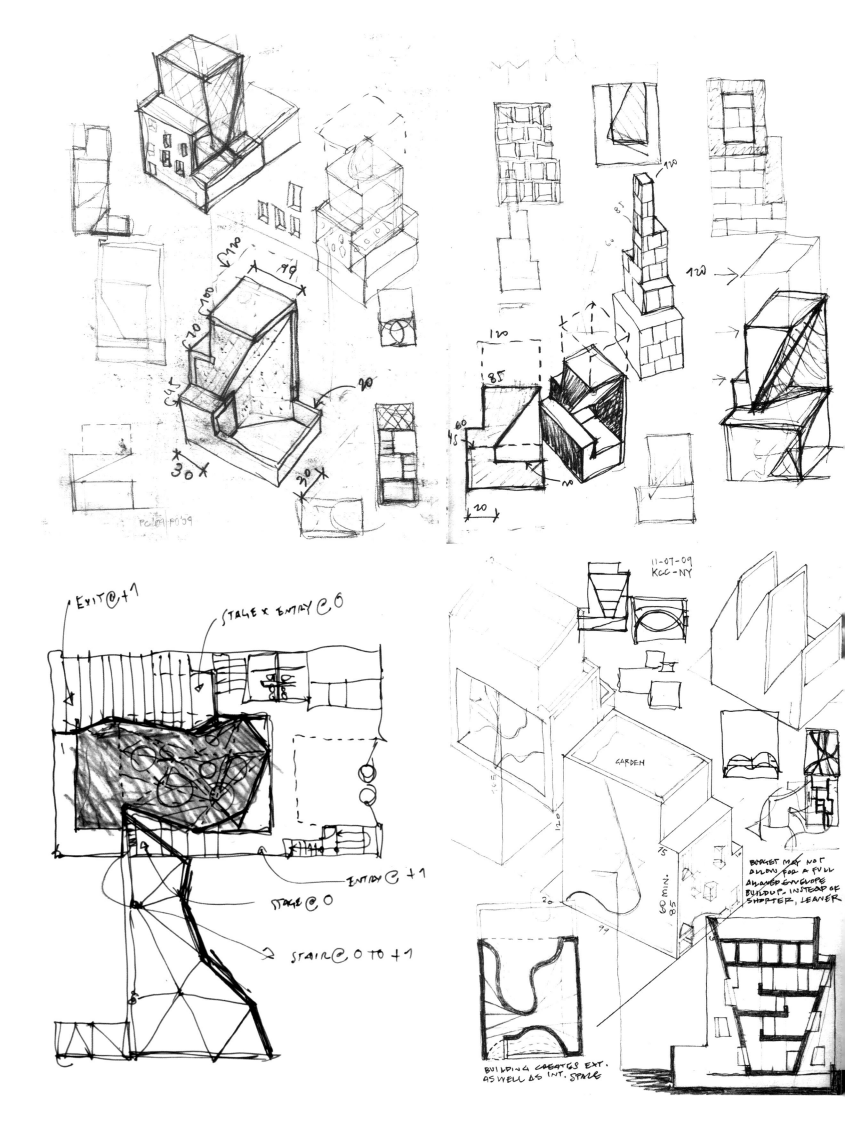

"Game Cube",
Hieroglyph,
Division without
Separation

"魔方"
象形文字
分而不离

韩国釜山歌剧院
Busan Opera House

设计单位：TOTEMENT / PAPER
项目地址：韩国釜山
设计团队：Levon Ayrapetov
　　　　　Valeriya Preobrazhenskaya
　　　　　Egor Legkov
　　　　　Adelina Rivkina

Designed by: TOTEMENT / PAPER
Location: Busan, South Korea
Design Team: Levon Ayrapetov,
Valeriya Preobrazhenskaya,
Egor Legkov, Adelina Rivkina

设计理念

设计以"分而不离"为基本理念展开设计,融合了多元化的元素,整合了简约的外部立方体形态和复杂的内部结构,平衡了空间的水平线条和竖向线条,最后呈现出来的是有着醒目的形态、新颖的形式、极端的个性的巨型标志。

设计特色

建筑的形态给人带来许多视觉联想:多面的水晶体、盛开的花朵、没有脚柱的宝塔,给人深刻的视觉感受。这一体量由多个简单的体块组合而成,经过精细的切割和巧妙的布置,才形成如此独特的形态。这一设计的独特美感不是源于单个结构的旋转,也不在于个体的数量和规模,而是由众多相似却又有些微差别的元素共同构成的壮观景象。

空间象形文字是"立方体"的一个基本构成元素,其外部覆盖着金属板,底部的小平面映衬出大地和水面,顶部的小平面将天空和太阳吸纳进来,宛若立体的棱镜,将周围的景致都纳入其中。

剧院高耸在天地之间,观剧台位于建筑的每一个楼层,对所有人开放。这里有着开放的芭蕾舞教室、排练室、餐厅、舞台设计师工作室、舞台、办公室和一些仓库设施,当人们从一个楼层进入另一个楼层,会惊奇地发现自己正处于一个"内部的室外空间",体验一次难忘的"魔方"之旅。

Legend
图例

1. Auditorium　　　　　　　　1. 礼堂
2. Stage　　　　　　　　　　　2. 舞台
3. Foyer　　　　　　　　　　　3. 门厅
4. Lounge Zone　　　　　　　 4. 休息区
5. Open Space　　　　　　　　5. 开放空间
6. Orchestra Rehearsal Room　6. 管弦乐队排练室
7. Ballet Rehearsal Room　　　7. 芭蕾舞排练室
8. Individual Dressing Room　　8. 个人更衣室
9. Group Dressing Room　　　 9. 团体更衣室
10. Costume Room　　　　　　10. 服装室
11. Make-up Room　　　　　　11. 化妆室
12. Storage Room　　　　　　 12. 储藏室
13. W.C.　　　　　　　　　　　13. 卫生间
14. Void above the Room　　　14. 房屋上方的空间

Plan Level + 37.800
平面层 +37.800

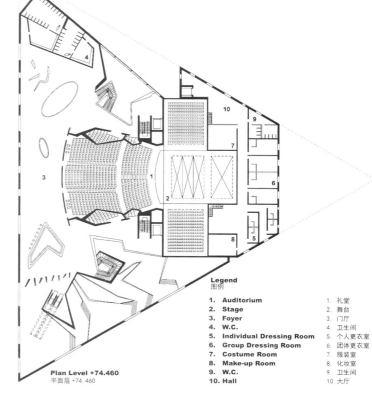

Legend
图例

1. Auditorium　　　　　　　　1. 礼堂
2. Stage　　　　　　　　　　　2. 舞台
3. Foyer　　　　　　　　　　　3. 门厅
4. W.C.　　　　　　　　　　　 4. 卫生间
5. Individual Dressing Room　　5. 个人更衣室
6. Group Dressing Room　　　 6. 团体更衣室
7. Costume Room　　　　　　 7. 服装室
8. Make-up Room　　　　　　 8. 化妆室
9. W.C.　　　　　　　　　　　 9. 卫生间
10. Hall　　　　　　　　　　　 10. 大厅

Plan Level +74.460
平面层 +74.460

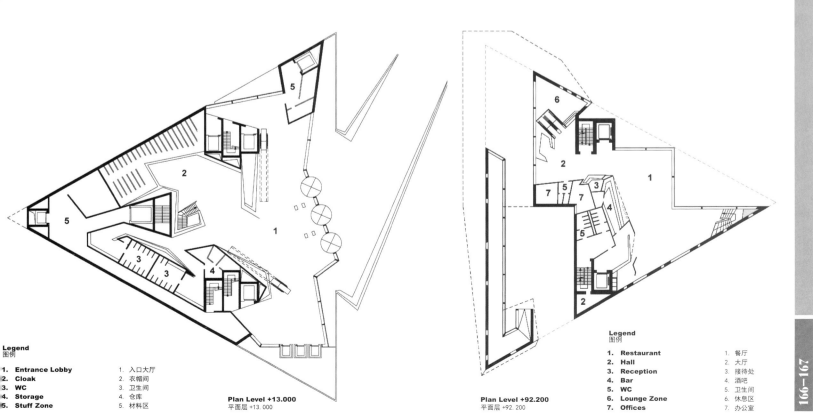

Legend
图例

1. Entrance Lobby — 1. 入口大厅
2. Cloak — 2. 衣帽间
3. WC — 3. 卫生间
4. Storage — 4. 仓库
5. Stuff Zone — 5. 材料区

Plan Level +13.000
平面层 +13.000

Legend
图例

1. Restaurant — 1. 餐厅
2. Hall — 2. 大厅
3. Reception — 3. 接待处
4. Bar — 4. 酒吧
5. WC — 5. 卫生间
6. Lounge Zone — 6. 休息区
7. Offices — 7. 办公室

Plan Level +92.200
平面层 +92.200

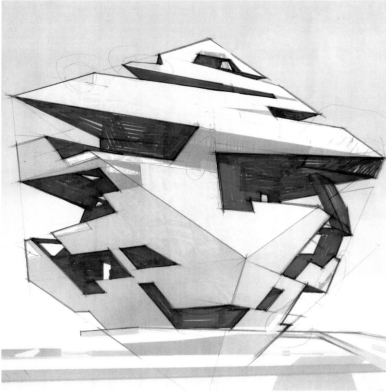
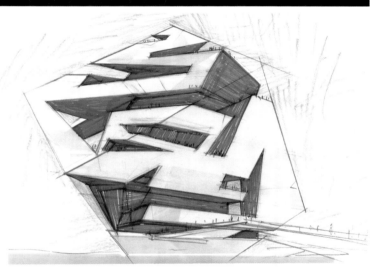

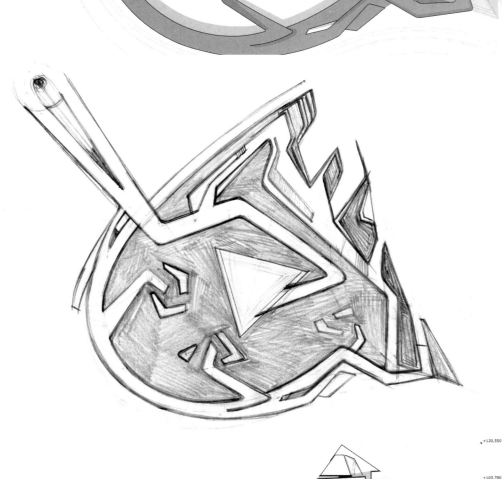

Design Concept

With the basic idea of "division without separation", the design has integrated diverse elements, merged elementary external cube-shape and complicated internal structure, and balanced horizontal and vertical lines in spaces to present a tremendous sign, a bright image, a new form and an extreme individuality.

Design Feature

The architectural form brings people plenty of visual associations: a faceted crystal, a blooming flower, a "pagoda without the socle" – allow a spectator to memorize an image of the building for long. The shaping concept of the buildings is based on placing simple forms together. The outside shape of the achieved agglomerate is made by cutting off the surpluses to make a cube form put on its edge. The esthetics of the given object depends neither on a separate turn of any of its parts nor on their size or quantity, but is defined by lots of similar but different elements with its external shape of an ideal Platon body in an unusual twist.

An elementary form of a cube – a spatial hieroglyph, is revetted outside by metal sheets, reflecting the Earth and the Water by its bottom facets, and the Sky and the Sun by its top ones. It's like a three-dimensional prism uniting in itself all the surrounding landscape.

Open stages – amphitheaters soaring between the Heaven and the Earth, observation decks are accessible to everyone. Some of ballet classes and rehearsal rooms, kitchens of restaurants and stage designers' studios, stages, offices and some warehouse facilities open over here. A visitor, travelling from level to level and finding himself on internally-external spaces, has an opportunity to observe the inner life of the building, enjoying an unforgettable journey inside the Game Cube.

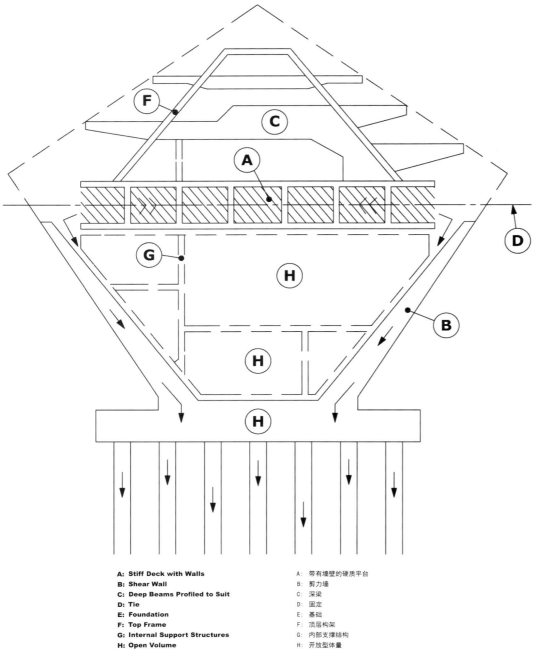

A: **Stiff Deck with Walls** A: 带有墙壁的硬质平台
B: **Shear Wall** B: 剪力墙
C: **Deep Beams Profiled to Suit** C: 深梁
D: **Tie** D: 固定
E: **Foundation** E: 基础
F: **Top Frame** F: 顶层构架
G: **Internal Support Structures** G: 内部支撑结构
H: **Open Volume** H: 开放型体量

Wind Turbine 风力涡轮机
Wind 风
Solar Panels 太阳能板
Thermal Energy 热能

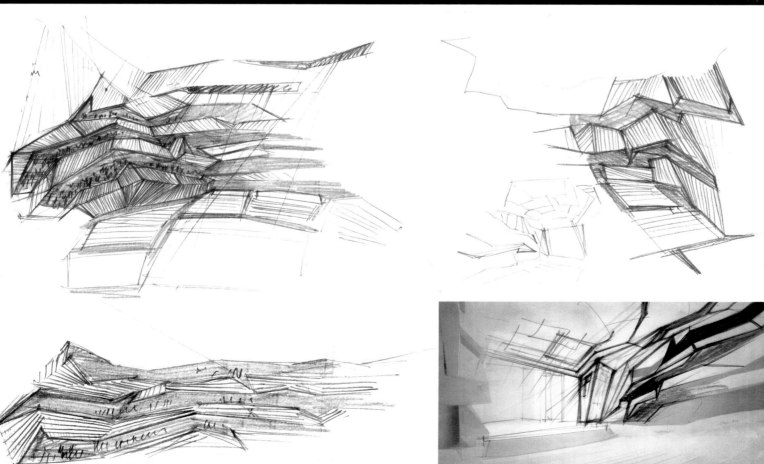

Pure Cloud,
Honeycomb
Rain Screen,
Split-level Design

至纯云彩
蜂窝雨幕
错层式设计

美国犹他州帕克城金博尔艺术中心
Kimball Art Center

设计单位：Brooks+Scarpa Architects	Designed by: Brooks + Scarpa Architects
合作单位：Blalock & Partners	Collaboration: Blalock & Partners
开发商：金博尔艺术中心	Client: Kimball Art Center
项目地址：美国犹他州帕克城	Location: Park City, Utah, USA
总建筑面积：3 279 ㎡（其中新增部分 2 071 ㎡，改造部分 1 208 ㎡）	Gross Floor Area: 3,279 m² (New: 2,071m², Renovation: 1,208 m²)
设计团队：Lawrence Scarpa　Mark Buckland　Angela Brooks　Silke Clemens　Jordon Gearhart　Ching Luk　Emily Hodgdon　Diane Thepkhounphithack　Amelia Wong　Kevin Blalock（Blalock & Partners）	Design Team: Lawrence Scarpa, Mark Buckland, Angela Brooks, Silke Clemens, Jordon Gearhart, Ching Luk, Emily Hodgdon, Diane Thepkhounphithack, Amelia Wong; Kevin Blalock (Blalock & Partners)

设计构思

除却近 36 421 700 ㎡的滑雪场、64 座载入国家史迹名录的历史建筑，帕克城最让人惊叹和着迷的当数无垠的深蓝色天空，那片无垠、纯净的天空，如美妙的梦境缠绕在游客心中，也成为本方案的灵感来源。为摆脱原有建筑的笨重体量，新的金博尔艺术中心的扩建部分试图从感官上将帕克城天空的独特性直接融入城市中，赋予建筑轻灵、飘逸的美感。

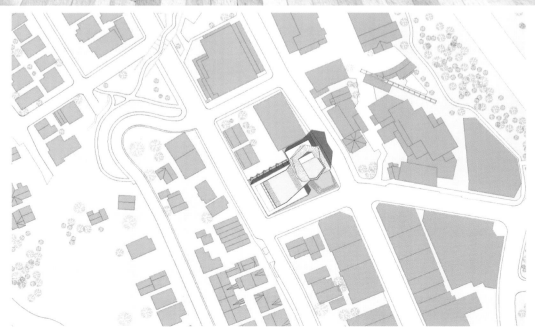

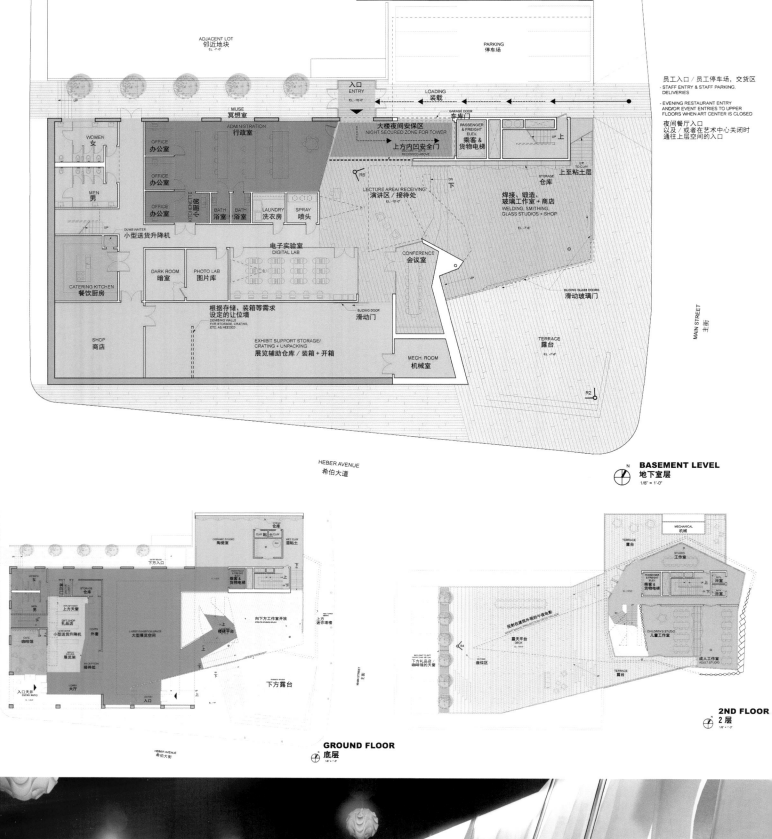

BASEMENT LEVEL
地下室层

GROUND FLOOR
底层

2ND FLOOR
2层

设计特色

新建建筑的上层用玻璃系统搭建，覆盖了透明的蜂窝状雨幕，底层立面采用透明玻璃，直接向街道开敞，并与金博尔艺术中心原有的笨重体量巧妙地交织在一起。最终形成的建筑体量，较低的楼层融入城市环境和周围已有建筑中，较高的楼层则悬于透明的空间之上，好似一朵飘浮的"云彩"。

在建筑内部，设计旨在将艺术空间巧妙地连接在一起，使当地的人们能自由地参与其中，不仅能欣赏艺术品，还能观看艺术创作的过程，享受到一种别样的艺术体验。设计将位于缅因街和赫伯大街拐角处的大型户外庭院与艺术家的工作室建立直接联系，户外庭院因此成为工作室的延展空间。

从庭院衍生出来的空间深入建筑内部，将新建建筑与已有建筑联系起来。这种错层式的设计既确保了空间的灵活性，使文化中心可根据不同的功能和需求对底层空间进行重新规划，也保持了视觉上的开放性和整体性。

在错层式底层的上方是附加的三个楼层和一个屋顶平台。建筑的第四层，也是最上层空间是为餐厅配备的租赁空间，从这里可以俯瞰整个城市及其周边的区域。餐厅的上方是一个公共屋顶平台，可作为各种类型公共和私人活动的场所。

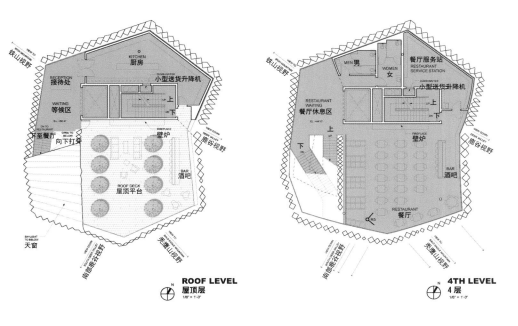

EXTERIOR SPACES
外部空间

All Public Courtyards and Plazas Receive Sunlight throughout the Day
全天所有公共庭院和广场接收到日光

- Roof Decks 屋顶平台
- Street Level 街道水平面
- Public Plaza 公共广场

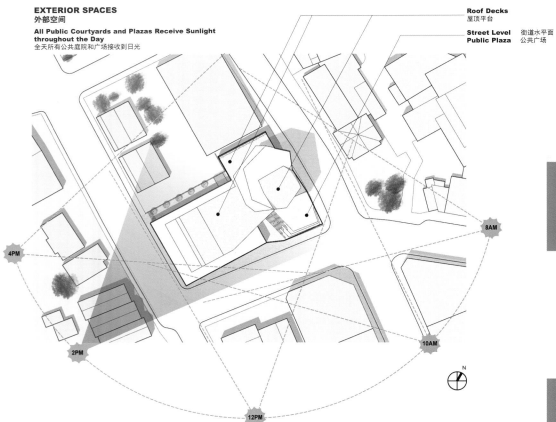

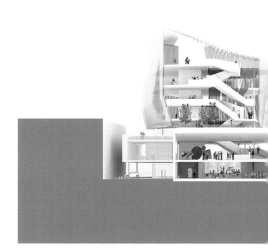

1. PHASE I – Initial
Isolate Kiln and Existing Stair/Elevator Core
1. 阶段一——初始
分隔窑炉和原有楼梯／电梯核心

2. PHASE I – In Progress
Temporary Stair Connecting Historic Building and New Wing Partial Construction of Final Plaza Steps Kiln in New Wing Becomes Operational
2. 阶段二——正在进行中
临时阶梯与历史建筑和新翼楼部分建筑广场台阶相连
新翼楼内的窑炉可操作

3. PHASE I – Near Completion
Completion of Plaza Stairs and Entrance through Historic Building
3. 阶段——即将完成
完成广场台阶和历史建筑的入口

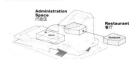
4. PHASE II
Administration Space Returns to Historic Building
Restaurant to Occupy Former Admin Space in New Wing
4. 阶段二
行政空间回到历史建筑
餐厅占据新翼楼原来的行政空间

1. EXISTING
Historic Building
1. 现存的
历史建筑

2. EXTEND ACROSS
Maximize Site Utilization
2. 横跨
最大化场地利用

3. EXTEND UP
Expand Usable Space on Footprint
3. 向上扩展
扩展可用空间

4. CARVE
Add Public Space at Street Corner and Parking under New Addition
4. 切开
在街角增加公共空间，在新的附加建筑加停车空间

5. CONNECT
Link Existing and New Building with Rooftop Deck
5. 连接
屋顶平台连接原有建筑部分和新建筑部分

6. SLICE
Maximize Usable Space within New Building
6. 切片
最大化新建筑内的可用空间

7. CONNECT
Augment Floor Plates to Suit Program, Environmental Factors, and Urban Context
7. 连接
扩充楼面板以适应规划、环境因素和城市环境

8. WRAP
Clad New Building in Light, Ethereal Skin
8. 包裹
以轻盈表皮包裹新建筑

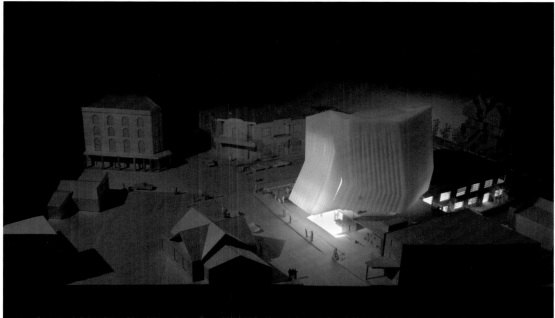

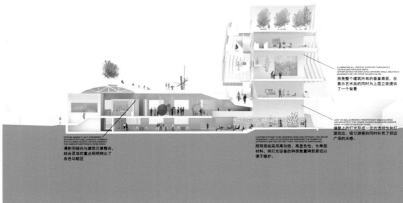

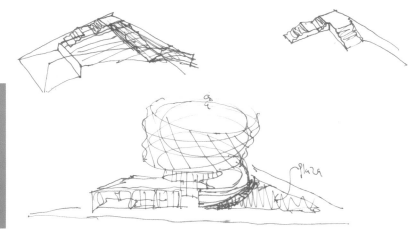

Design Concept

Besides over 9,000 skiable acres and 64 buildings on the National Register of Historic Places, one of the most incredible and mesmerizing natural features of Park City is the seemingly endless deep blue sky. The sky always seems to quickly return to its infinite and hypnotic clarity, with rarely with a cloud in the sky, provoking a dreamlike state of mind of viewers. To distinguish from the heavy mass of the existing historic Kimball buildings around, the concept for the new Kimball Art Center addition and renovation is to perceptually bring the uniqueness of the Park City sky directly into the city.

Design Feature

The upper floors are composed of a conventional glazing system that is covered by a rain screen made from a more translucent honeycomb material. The new ground level façade is constructed of very transparent glass and opens directly to the street, while delicately connecting and weaving into the heavy mass of the existing historic Kimball building. As a result, the lower floor is absorbed into the context of the city and the adjacent existing building, while the upper floors overhang the more transparent level below, like a floating "Cloud".

Inside the building, the design intent is about delicately knitting together art spaces, allowing the community to view and participate in the artistic experience. Rather than simply displaying art for view, the new design reveals to the community the very process by which art is created. A large exterior court at the corner of Main Street and Herber Ave. links directly to the studios, which would become an outdoor workspace.

These studio spaces flow from the court deep into the building, linking the new structure with the existing. This split-level design provide the ability to easily divide and use the ground level for a variety of purposes and functions, while remaining visually open and whole.

Above the split-level ground floor, there are three additional floors and a roof deck. The fourth and uppermost floor is a leasable space for a restaurant with commanding views overlooking the city and into the surrounding areas. Above the restaurant, the building culminates with a public roof terrace, that can be used for a variety of public and private events.

韩国大邱高山郡公共图书馆
Daego Gosan Public Library

设计单位：Process-based Architecture Studio
开发单位：大邱广域市寿城区办事处
项目地址：韩国大邱
设计团队：Jafar Bazzaz
　　　　　Arash Pouresmaeil
　　　　　Kamal Youssefpoor

Designed by: Process-based Architecture Studio
Client: Daegu Metropolitan City Suseong-gu Office
Location: Daegu, South Korea
Design Team: Jafar Bazzaz, Arash Pouresmaeil, Kamal Youssefpoor

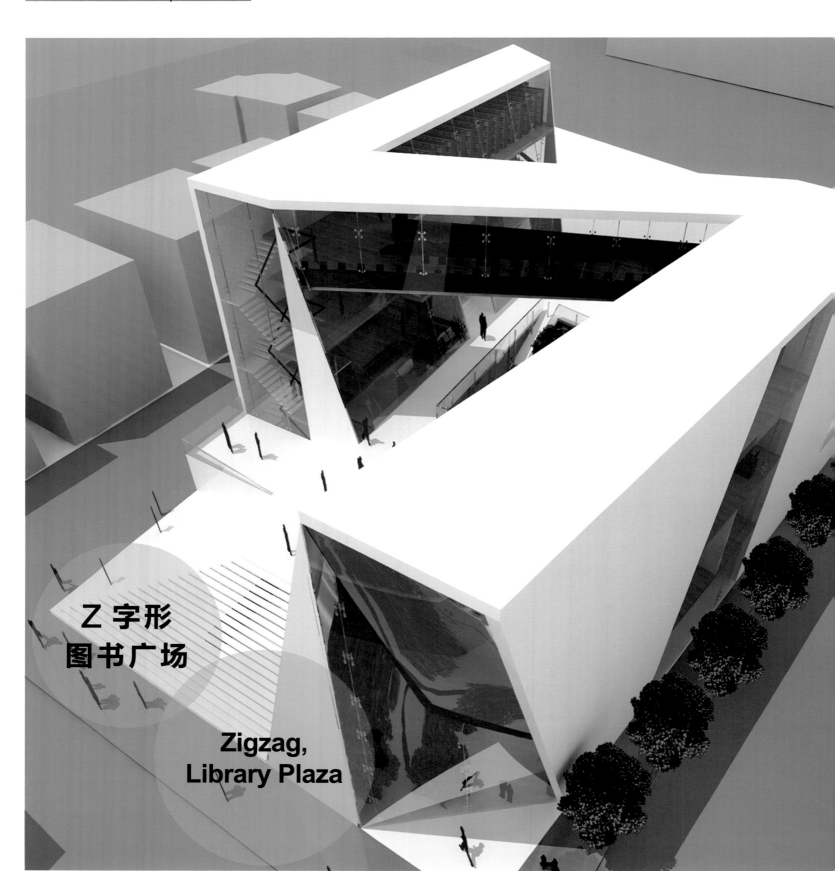

Z 字形
图书广场

Zigzag,
Library Plaza

设计构思

设计以营造一个充满生机和活力的城市空间为理念而展开，以建立图书馆和城市之间的直接联系。故在构思过程中，除了构建图书馆大楼，设计师还设置了一个面向公园和城市的城市广场，因此，这一项目被生动地称为"图书广场"。

设计特色

图书馆包括综合性的建筑体量和两栋独立的翼楼。两栋翼楼由一座"桥梁"衔接，使整体结构呈"Z"字形，"桥梁"内容纳了图书馆的综合阅览室。在两栋翼楼之间、地面层的上方，有一个面向城市的广场，大面积的楼梯将之与城市空间连接起来。

图书馆广场提供了一个面向公园和城市的大型平台，以充当图书馆的中央核心，其较低的部分是一个有顶的广场，由大厅和公共空间组成，两个广场通过一个下沉花园连接。这一空间既是一个可散步和社交的开放空间，同时也是一个静谧的学习空间。

Design Concept

The project design is based on the concept of creating a lively and dynamic urban space and building a direct connection between the library and the city. Besides the library building, designers have envisaged an urban plaza facing the park and the city which is called the "Book Plaza".

Design Feature

The library consists of an integrated ground level and two separated wings. These wings are connected together through a bridge that includes the general reading room of the library. In between of these wings and over the ground level, there is a plaza facing the urban context that is connected to the city through extensive stairs.

The library plaza provides a big terrace towards the park and the city. It acts as the central core of the library and its lower space is a roofed plaza that consists of lobby and common spaces. The roofed plaza and the main plaza are connected together through a sunken garden. The library provides an open public space in the plaza for walking and social interactions, and private study spaces facing the plaza.

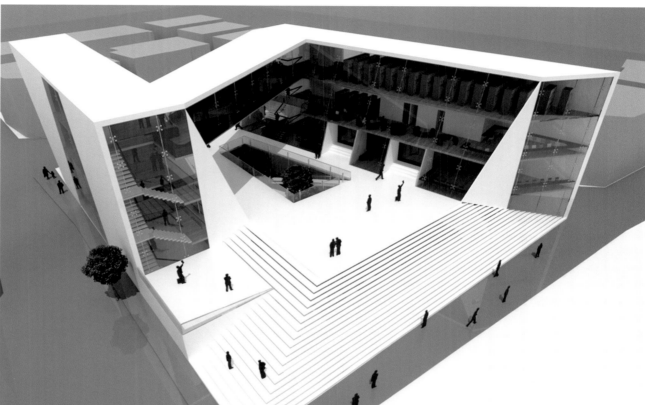

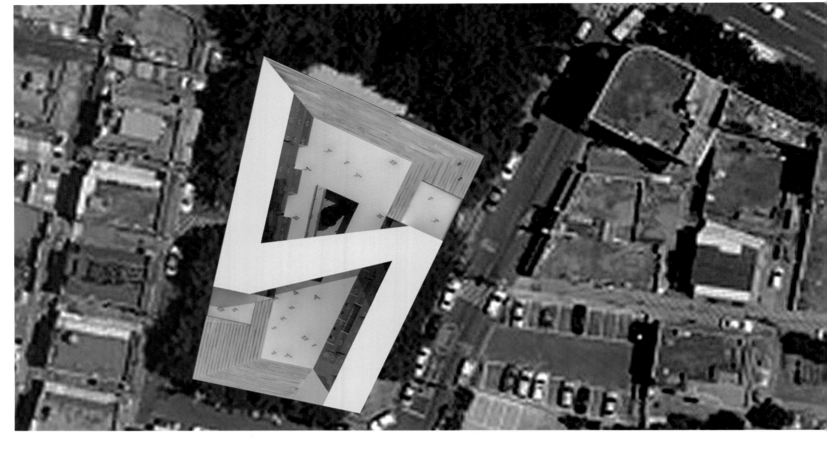

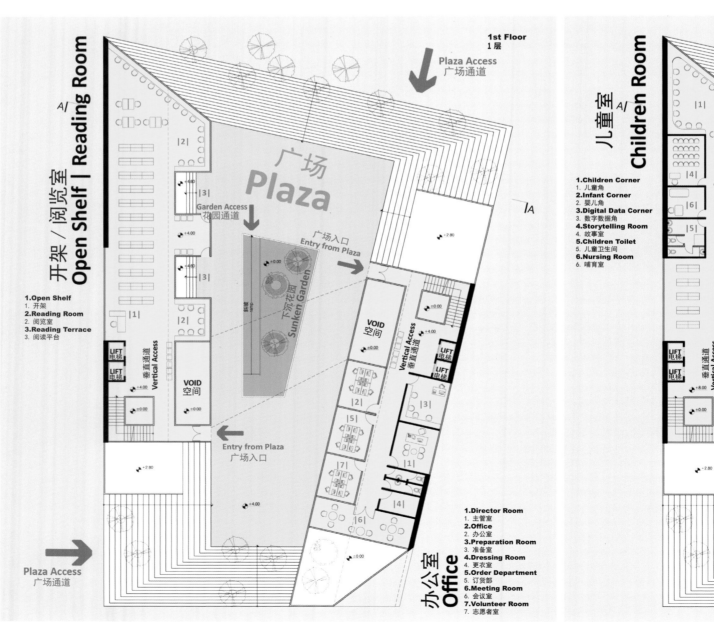

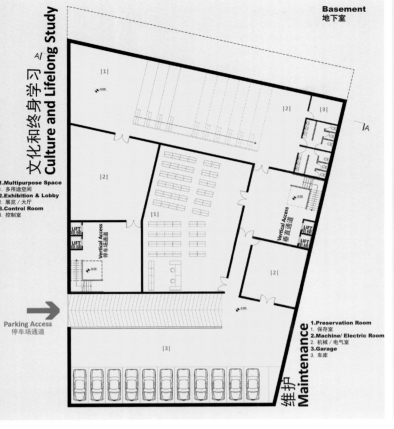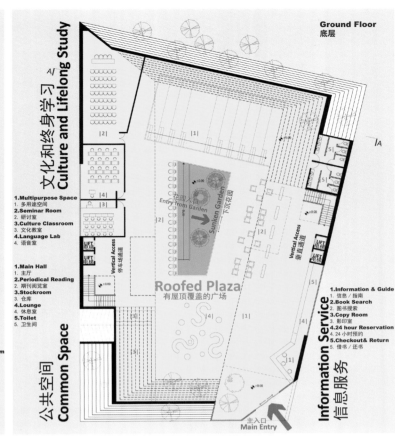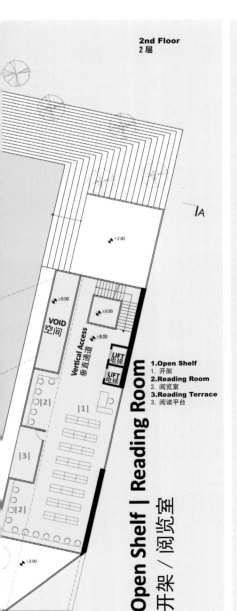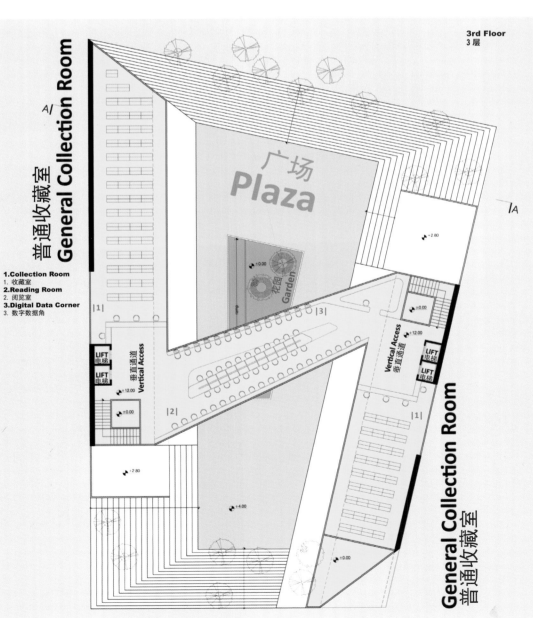

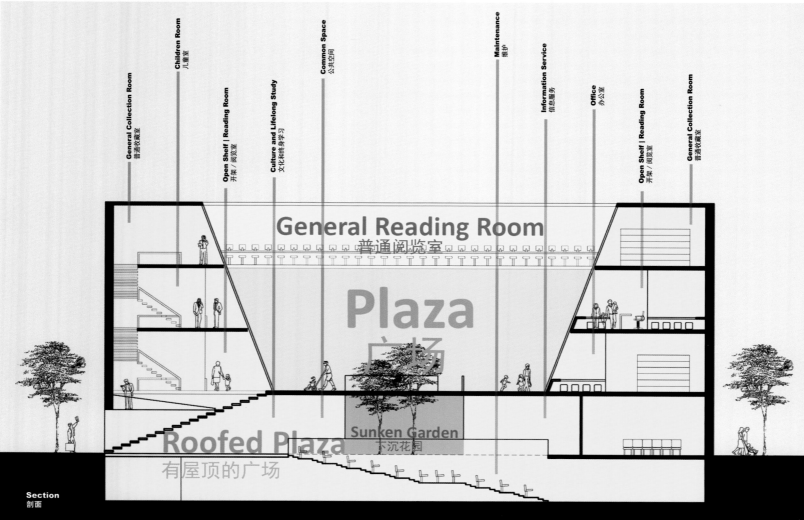
Section
剖面

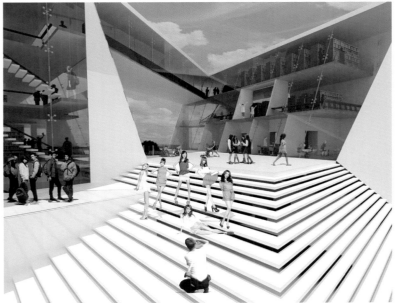

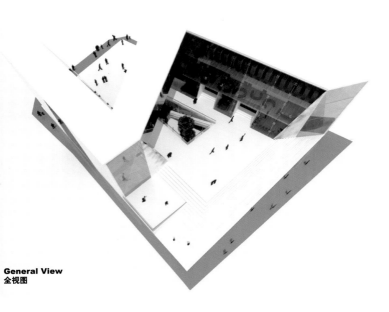

General View
全视图

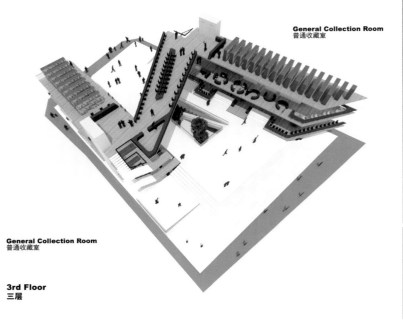

General Collection Room
普通收藏室

3rd Floor
三层

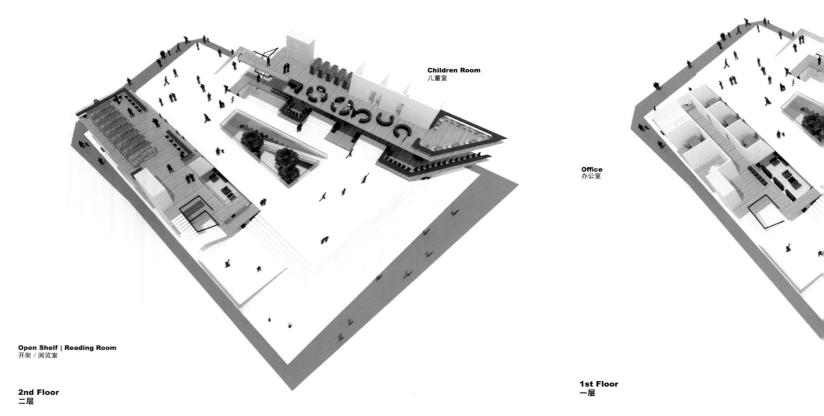

Open Shelf | Reading Room
开架 / 阅览室

Children Room
儿童室

Office
办公室

2nd Floor
二层

1st Floor
一层

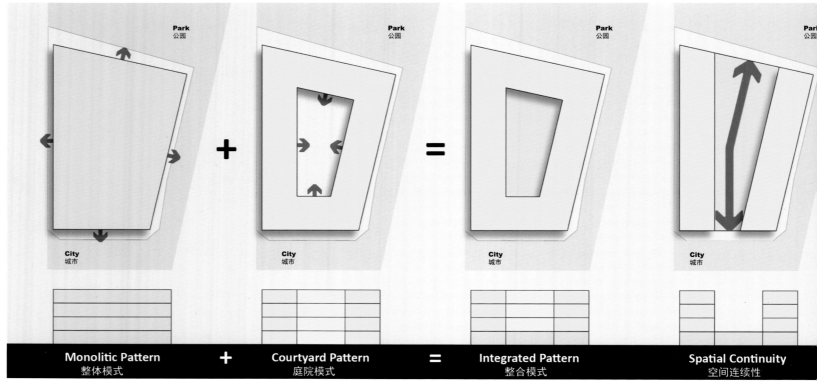

| Monolitic Pattern | + | Courtyard Pattern | = | Integrated Pattern | Spatial Continuity |
| 整体模式 | | 庭院模式 | | 整合模式 | 空间连续性 |

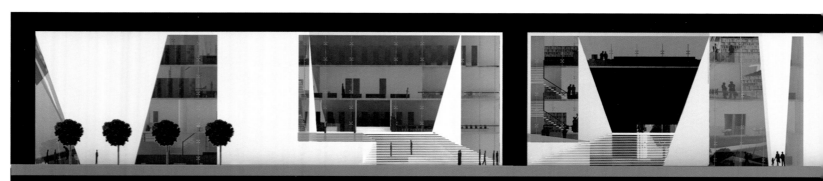

East Elevation
东立面

South Elevation
南立面

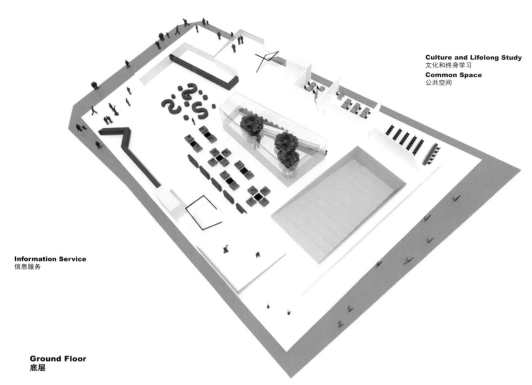

Open Shelf | Reading Room
开架／阅览室

Culture and Lifelong Study
文化和终身学习
Common Space
公共空间

Information Service
信息服务

Ground Floor
底层

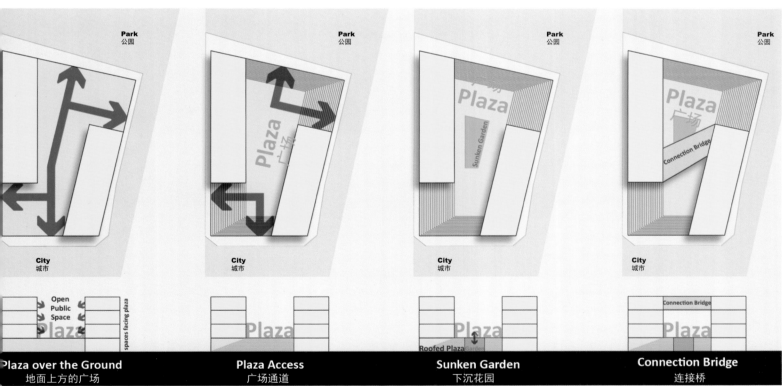

Plaza over the Ground 地面上方的广场　　**Plaza Access** 广场通道　　**Sunken Garden** 下沉花园　　**Connection Bridge** 连接桥

West Elevation 西立面　　**North Elevation** 北立面

俄罗斯南萨哈林斯克商业和展览中心
Business & Exhibition Center in Yuzhno-Sakhalinsk

设计单位：TOTEMENT / PAPER
项目地址：俄罗斯南萨哈林斯克
基地面积：2 000 m²
建筑面积：1 800 m²
设计团队：Levon Ayrapetov Valeriya Preobrazhenskaya
　　　　　Yegor Legkov Adelina Rivkina
　　　　　Darya Samohvalova Yevgeny Kostsov

Designed by: TOTEMENT / PAPER
Location: Yuzhno-Sakhalinsk, Russia
Site Area: 2,000 m²
Building Area: 1,800 m²
Design Team: Levon Ayrapetov, Valeriya Preobrazhenskaya,
Yegor Legkov, Adelina Rivkina,
Darya Samohvalova, Yevgeny Kostsov

方案一　Proposal I

设计构思

应客户要求,设计以"设计一个环境友好型的建筑"为主要理念来展开,这在理论层面上,已经超越了为了未来有资源可用而在脑子里形成的限量意识的资源保护,而是上升到环境保护的层面。

这是为南萨哈林斯克地区的一家能源公司设计的商业和展览中心,因此,在选址方面,设计综合考虑了诸多因素:首先,建筑要远离主要的生产区,即远离防爆危险区,以保证建筑的安全性;其次,考虑到交通的便捷性以及该工厂对出入人员的严格管理制度,设计师将建筑靠近道路布局,以优化交通设计方案。

设计特色

该方案是对新型建筑的一种探索。建筑的内部结构由简单的元素构成,这些元素的规模和大小依据建筑的功能来决定,而外部形态则构成了一个简洁的平行六面体。建筑的内外结构虽不同,却能产生相互作用:首先,建筑的外部结构是由内部元素来定义的,这些元素经过严格的"程式化"后,才构成这个简约的外部形态;其次,外部形态作为内部结构的一个剖切面,赋予了内部结构新的活力。

设计师以圆锥体为基本空间形态,这一形态体现了向心力和离心力的动态旋转。设计师将这些中空的圆锥体以不同的方式插入到体量中,在建筑内部形成了各空间中的连接通道,使空间与空间之间隔而不离。这种兼顾水平和垂直方向上的连接方式,成为设计师对不同形态空间之间相互作用的探索。

Cone, Parallelepiped, Varied Forms

**圆锥体
平行六面体
各异形态**

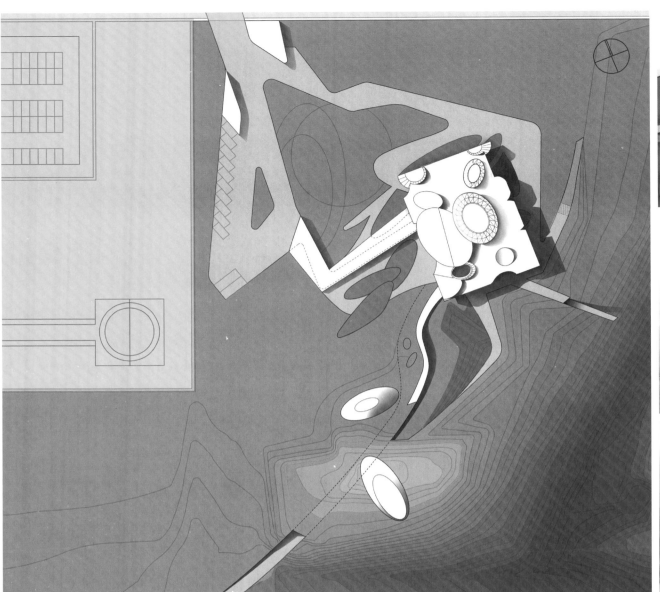

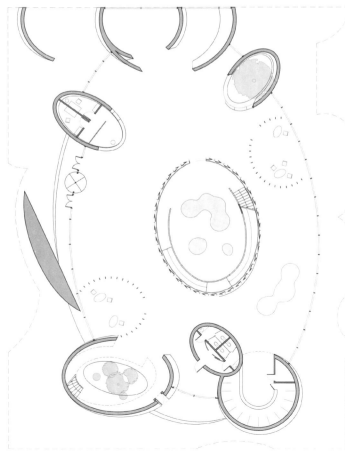

First Floor Plan
一层平面图

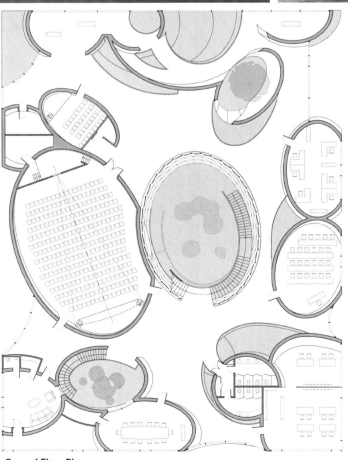

Second Floor Plan
二层平面图

Design Concept

The design concept is based on the client's requirement of designing a building friendly to the environment. It has surpassed the awareness of "protection of resources for future use", but reached to the level of environment protection.

The project is a business and exhibition center for a energy company in Yuzhno-Sakhalinsk. A construction site was chosen according to a great number of conditions: first, its location should be far from the main production building in order to keep away from the ex-zone; second, it should be close to the road in order to optimize traffic scheme and access for visitors since permission to be present in any part of the plant is strictly specified.

Design Feature

The scheme provides an opportunity to look for a new type of building. Its inner structure is composed of simple elements with different scales depending on inside or outside function with its outside cover made as another simple form – a parallelepiped. The inside and the outside interact with each other despite of their structural differences: the outside form is defined by internal elements. All elements, after being strictly "programmed", form this simple external form; the outside form serves as a cutting plane and gives a new life to the inside one.

Taking cone as a basic form, the building form reflects a dynamic rotation, both centripetal and centrifugal. The "vacuum" cone penetrating to a mass in different ways creates connecting channels in internal spaces, which makes the spaces divided but not separated. The vertical and horizontal connections become an occasion for designers to keep on researching on various forms and their interactions.

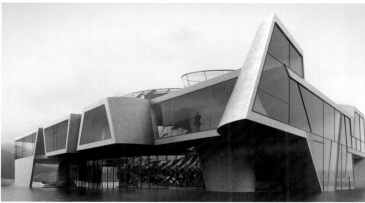

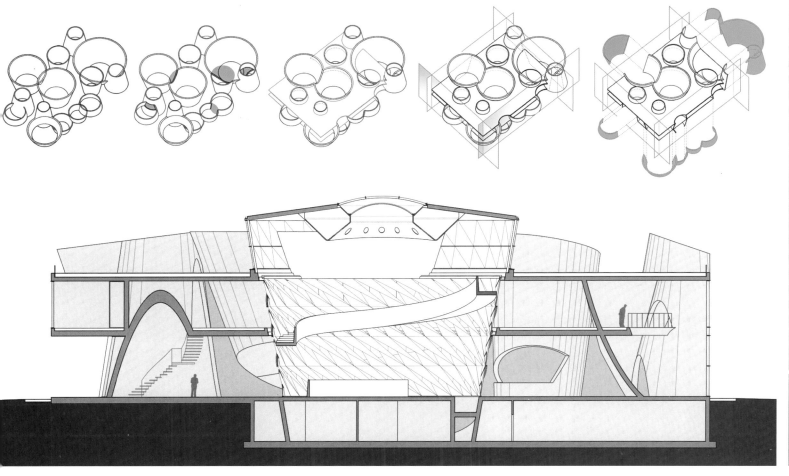

方案二 Proposal II

无立面
残缺美
动感镜头

None Façade,
Beauty of
Defective,
Dynamic Shot

设计特色

不规则的建筑形态，使建筑并没有主要的立面或是其他的外立面，或者说根本没有概念意义上的立面，它所要表达的是"休息中的运动"呈现出来的美感：水平延展开来的建筑，似即将展翅飞翔，又似静卧在山间休憩，动与静交织在一起，产生一种独特的美。

建筑是一个连续的主体，其形态只是一个中间状态，或者说是无数形态中的一个"镜头"，这些"镜头"都很类似，但又各有不同，主要是通过固态的金属、混凝土和玻璃来展现这动的动感。空间的通透性与不完整性、建筑与景观之间的呼应，以及倾斜的窗体和彩色玻璃窗体，这些元素都表达了东南亚和东亚文化艺术中永恒的缺陷美。

Design Feature

The structure of the building is made in such a way without main and additional façades and without a façade as a notion at all. It is based on the principle of "motion at rest": the horizontal stretching building seems to start to fly or sometimes hanging rest on the hills, exerting unique beauty between the dynamic and the static.

The building is a body and its shape is just an intermediate stop or one "shot" of millions, very much alike, but always different when the main aim is to reflect dynamics of motion with the help of the mass of frozen metal, concrete and glass. Transparency, incompleteness, reference to the landscape, uncommon angles of windows and stained-glassed windows – all these brings us to the everlasting incompleteness of South-East and Far East Asia art.

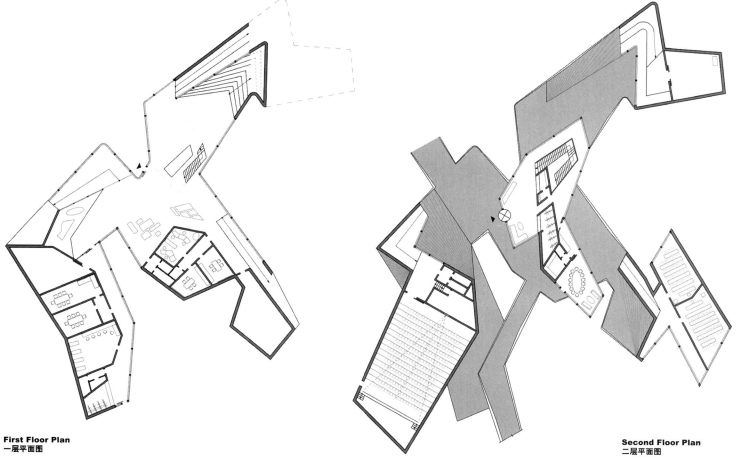

First Floor Plan
一层平面图

Second Floor Plan
二层平面图

	Halls	大厅
	Exhibition Spaces	展览空间
	Cloak Room	衣帽间
	Cinema Hall	电影院大厅
	VIP	VIP室
	Meeting Halls	会议厅
	Utility Rooms	杂物间
	Computer Display Room	计算机展示室
	LNG Display Area	LNG展示区
	W/C Rooms	W/C室
	Storage Room	储藏室
	Technical Rooms	技术室
	Operation Room	操作室
	Offices	办公室

重庆"城市森林"
Chongqing "Urban Forest"

设计单位：MAD 建筑事务所	
合作单位：奥雅纳工程顾问公司	
项目地址：中国重庆市	
基地面积：7 700 ㎡	
建筑面积：216 000 ㎡	
设计团队：马岩松　党　群	
于　魁　Diego Perez	
赵　伟　Chie Fuyuki	
傅昌瑞　Jtravis B Russett	
戴　璞　Irmgard Reiter	
覃立超　Rasmus Palmqvist	
谢新宇	

Designed by: MAD Architects
Collaboration: ARUP Group Ltd
Location: Chongqing, China
Site Area: 7,700 m²
Building Area: 216,000 m²
Design Team: Ma Yansong, Dang Qun, Yu Kui, Diego Perez, Zhao Wei, Chie Fuyuki, Fu Changrui, Jtravis B Russett, Dai Pu, Irmgard Reiter, Rasmus Palmqvist, Qin Lichao, Xie Xinyu

Vertiginous Hillsides, Cave Space, Urban Forest

蜿蜒山脉
巢穴空间
城市森林

设计构思

当高密度的水泥丛林急剧在城市中扩张开来,当自然逐渐在人们眼前消失,越来越多的人开始思索"城市的未来"这一问题。MAD 认为,推动城市化发展进程的不仅仅是经济的增长和物质的繁荣,更多的是引导城市灵魂的文化。

在对中国经济发展现状、社会环境和全球化大背景这些因素进行综合考量后,MAD 在重庆这座山水之城的市中心设计了一座高 385m 的城市文化综合体——"城市森林"。这是针对中国城市发展进程而提出的新建筑概念,旨在在中国最年轻的直辖市创造绿色立体城市,让自然重新融入未来的高密度城市,同时唤起曾经寄托于东方山水之间的天然情怀。

设计特色

设计师汲取了东方哲学中对"自然"和"人造"的见解,将现代城市生活与自然山水体验结合起来,设想了这座"城市森林"。设计师的灵感来自于国画中的山水景观,建筑的形体如山脉般蜿蜒起伏,生动、流畅而富于变化,宛若自然的延续。

这座高耸的大楼每一层都是抽象的曲线形,大楼的中心是一个圆柱体结构,其外是由玻璃包裹的墙体结构,这种设计既加强了建筑的透明性,同时也使每层楼看起来都好似悬浮在其下楼层之上。

"城市森林"不再强调垂直的力量,而是更加注重人们在多向度空间的体验和感受:多层的立体花园、浮游的平台、纯净光洁的巢穴空间,建筑的形式消失在空气、风和光线的空间流动之中,置身于其中,人们将与自然不期而遇。

这种将东方人文主义的自然精神与城市公共空间结合的城市发展模式,将成为绿色立体城市的先驱,"城市森林"不再是冷硬的"城市机器",而是一座在钢铁混凝土林立的城市中心内自然呼吸的人造有机体。

Design Concept

As high density concrete clusters radically expanding and the natural environment gradually vanishing in cities, more people get to worry about the cities' future. MAD believes, urbanization at the macro scale is intended to drive economic prosperity, but it must also engender a cultural identity for this newly announced megacity in the hinterlands.

Based on a comprehensive consideration of China's current economic development, social environment and the background of globalization, MAD designed a 385-meter cultural complex – "Urban Forest" in the landscape city Chongqing. The new architectural concept of this project targets to the urban development of China, which strives to create a green vertical city for the youngest urban center of China, Chongqing. It is a tower that reincorporates nature into a high-density urban environment and evokes affection for nature now lost in modern global cities.

Design Feature

The Urban Forest draws inspiration from the appreciation of nature and the artificial in oriental philosophy and reconnects urbanity to the natural realm. The shape of the tower mimics vertiginous hillsides, shifting in a dynamic yet holistic rhythm.

Each floor of the tower has an abstract curved shape. A cylinder in the middle of the building covered by glass enhances the transparency of the building whilst makes each floor "floating" above the one beneath it.

Unlike its predecessors, the Urban Forest is a tower with no emphasis on machined vertical force. It concentrates on the multidimensional relationships within complex spaces: multistory sky gardens, floating patios and serene gathering spaces. Architectural form dissolves into the ephemeral movement of air, wind and light.

The urban development mode of incorporating the nature spirit of the oriental humanism with urban public space will become a pioneer of green vertical cities. "Urban Forest" is no longer a cold "City Machine", but a living organ that breathes new life into the steel and concrete city.

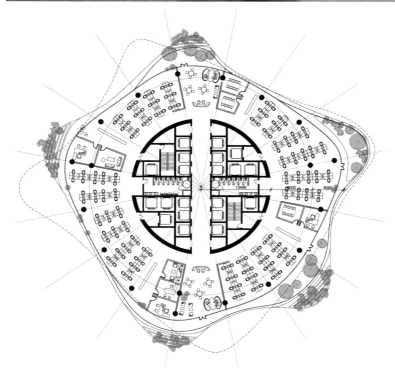

Open Office Plan
开放型办公室平面图

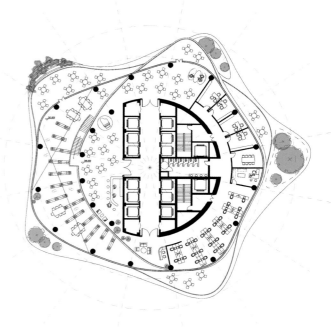

Duplex Office Plan
复式办公室平面图

City Center of Chongqing, 2008
2008年重庆市中心

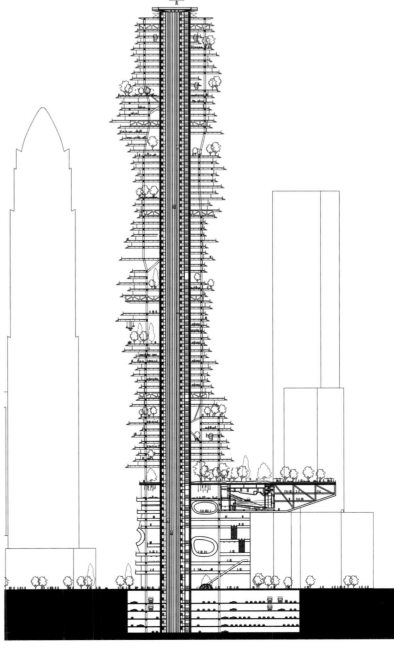

学校建筑
School Building

Test Field, Sloping Roof, Ecological Idea Model
试验田 倾斜屋顶 生态典范

丹麦奥尔胡斯纳维达斯公园工程学院和科学公园
Navitas Park Engineering College & Science Park

设计单位：C. F. Møller Architects
开发单位：奥尔胡斯工程学院
　　　　　奥尔胡斯海事和科技工程学院
　　　　　Incuba 科学公园
　　　　　奥尔胡斯市
项目地址：丹麦奥尔胡斯
项目面积：35 000 ㎡

Designed by: C.F. Møller
Client: Engineering College of Aarhus;
Aarhus School of Marine and Technical Engineering;
Incuba Science Park; Aarhus Municipality
Location: Aarhus, Denmark
Area: 35,000 m²

设计构思

设计师旨在将纳维达斯公园改造成为一个在新能源和环保技术领域走在世界前列的知识中心和创意园区，为学校师生、科研工作者和企事业单位提供一个在教育、研究、创业发展及能力应用等方面强有力的学术中心。

设计特色

倾斜的屋顶统一了建筑的整体风格，并赋予了建筑新的特色，是本方案的独特之处。屋顶本身就是一处独特的景观，并设计有庭院、观景台和学术试验田，这些元素将建筑完美地融入周围景观中，使之成为一栋富有生机和活力的建筑。

设计非常关注能源的利用及建筑的可持续性等问题。新建的学校建筑能耗低，建筑的隔热装置、太阳能电池、依据水的温差原理设计的热泵等设备的采用以及自然能源的使用，将使建筑成为先进的建筑群落典范。

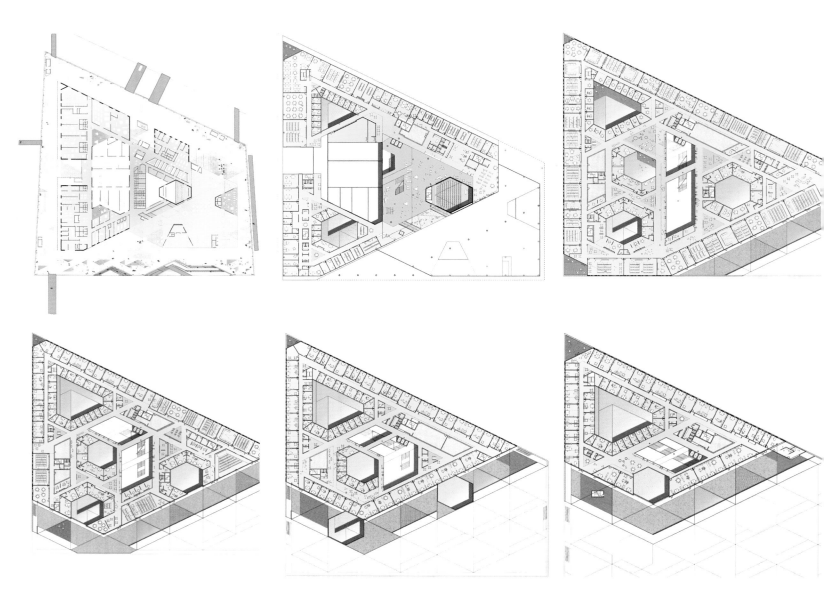

Design Concept

Designers aim to build a knowledge center and innovative environment primarily in energy, environment and building construction, an internationally competitive and attractive environment for students, teachers, scientists and companies, and a strong center for the development and application of competencies in education, research and entrepreneurship.

Design Feature

A sloping roof is the distinctive element of the competition proposal. The roof unites and gives character to the new building. The roof is a landscape in itself, with integrated courtyards, terraces and test fields for education and research purposes. These elements integrate the project with surrounding landscape to create a lively building.

The project design pays close attention to the use of energy and sustainability. The new buildings will be classified as low-energy class 1, using increased insulation, solar cells and heat pumps which make use of the temperature difference, to make a model of advanced building clusters.

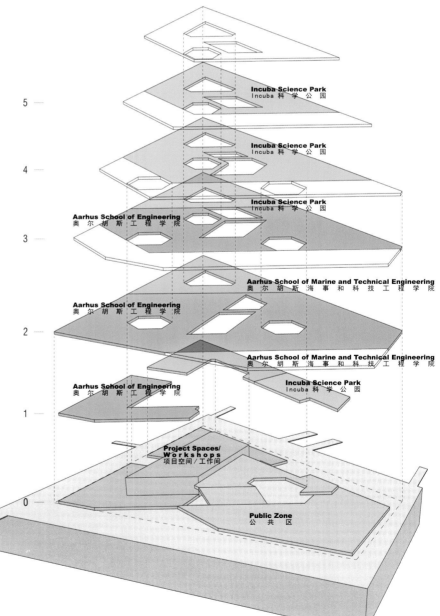

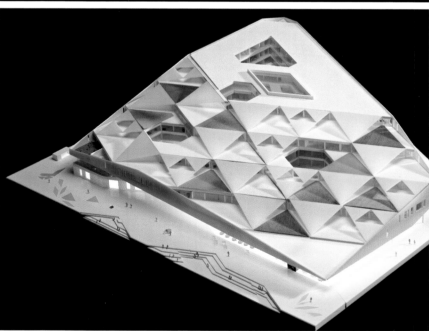

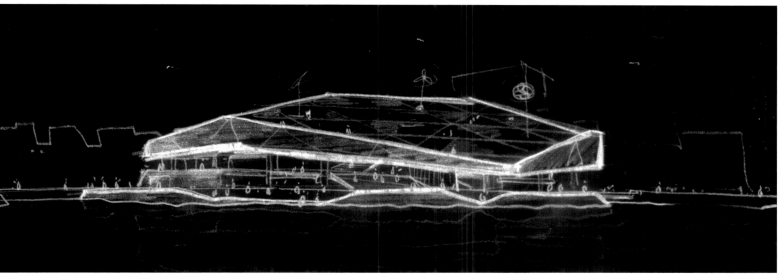

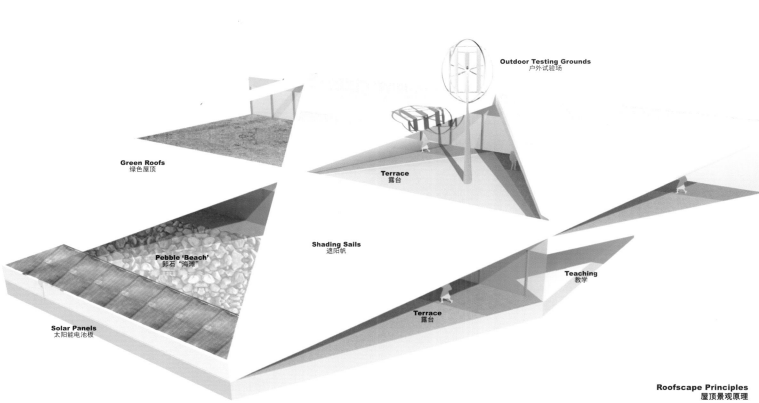

Roofscape Principles
屋顶景观原理

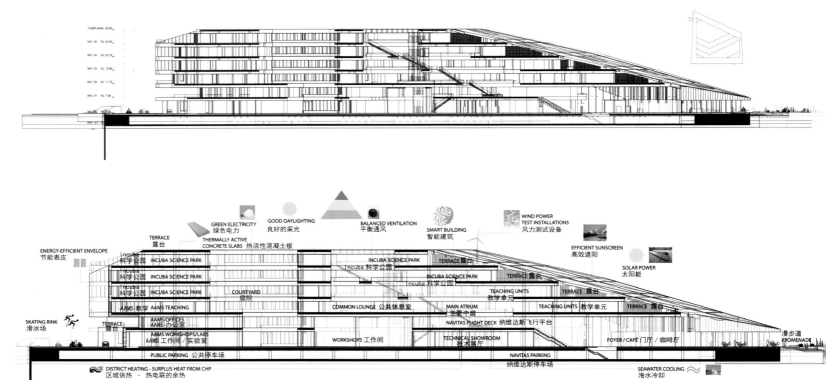

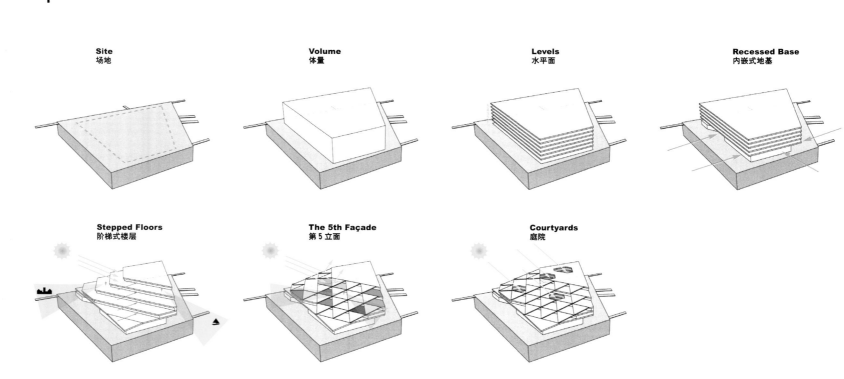

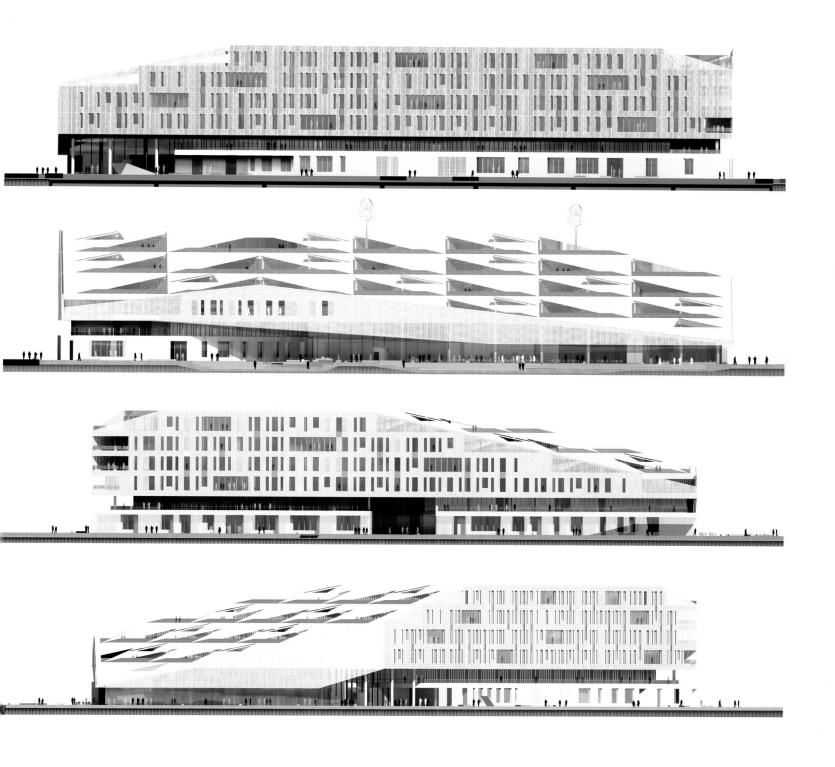

瑞士洛桑市洛桑联邦理工学院教学桥整修及机械学大厅和图书馆扩建

Teaching Bridge at the Ecole Polytechnique Fédérale de Lausanne/Rehabilitation – Extension of the Mechanics Hall and Library

设计单位：Dominique Perrault Architecture
开发单位：洛桑联邦理工学院
项目地址：瑞士洛桑市
基地面积：15 500 m²
建筑面积：5 100 m²

Designed by: Dominique Perrault Architecture
Client: EPFL
Location: Lausanne, Switzerland
Site Area: 15,500 m²
Built Area: 5,100 m²

设计构思

机械学大厅和原有的中央图书馆位于劳力士学习中心的上方，故本次改造和扩建方案在一定程度上延续了劳力士学习中心的建设风格。考虑到要将中心图书馆转移到这一新的旗舰式建筑内，如何对因图书馆搬迁而腾出来的空间进行合理安置，引发了设计师对学校中心服务区进行功能重组的思考。因此，设计师为工程和科学学院的学生新建了几个实验室，在劳力士学习中心和会议中心之间营造了一个活力空间，将原有的校园建筑与新建筑联系起来。新老建筑间自然而紧密的衔接关系，也是本方案脱颖而出的一个设计要点。

设计特色

在项目的第二期，设计师在校园的最高处，即皮卡尔大街的上方构建了一个由南至北穿过基地的宏大建筑体量。这个白色的体量构建在街道的上方，好似一座桥梁将两端的建筑群连接起来，因此，该建筑也被设计师形象地称为"教学桥"。

除了将这一异形的场地连成一个连续的整体，考虑到这个建筑是为展示多元的教学方式、满足多样化的教学目标而专门设计的，因此，这一建筑也为该校推行新的教学形式创造了条件。这种在校园中心以城市规划途径引导校园新动态的思维方式，也是备受客户推崇的要点。

Design Concept

Located just above the Rolex Learning Center, the mechanics hall and the old central library will be completely transformed and extended, following to a certain extent the construction of the Rolex Learning Center. The contents of the central library were transferred to this new flagship building of the school, thus liberating important volumes. Properly arranging vacated spaces to regroup a part of the school's central services and create several new laboratories for students of the Engineering and Science Faculties is paid close attention to. The architect created a new lively space located between the Rolex Learning Center and the congress center and also established a strong link between the old buildings of the campus and the new constructions in the South. The natural but close connection between the new and old buildings is a design highlight of this scheme.

Design Feature

In the second phase, an ambitious construction, the highest of the campus, could be built above Avenue Picard, which is crossing the site from south to north. The white volume above the avenue is called "Teaching Bridge" as it connects the building clusters of both ends.

The heterogeneous place is designed as a continuous whole. The project is designed to satisfy multiple teaching methods and goals, which creates conditions to put forward modern pedagogical forms. A true urban planning approach "creating a new dynamic in the heart of the campus" in this project is welcomed by the client.

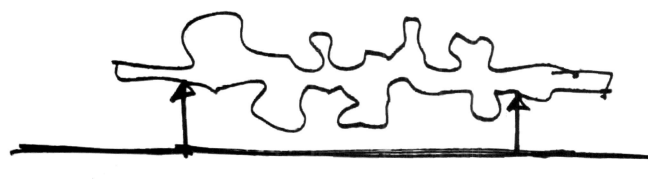

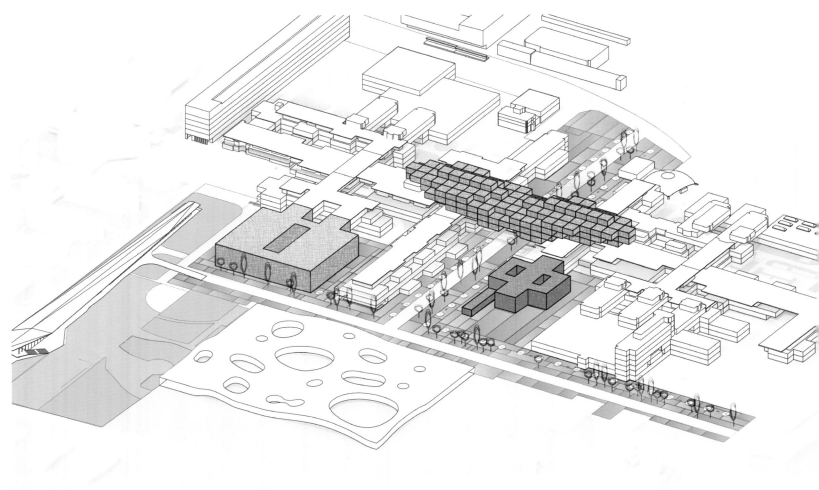

Urban Intention
城市意向

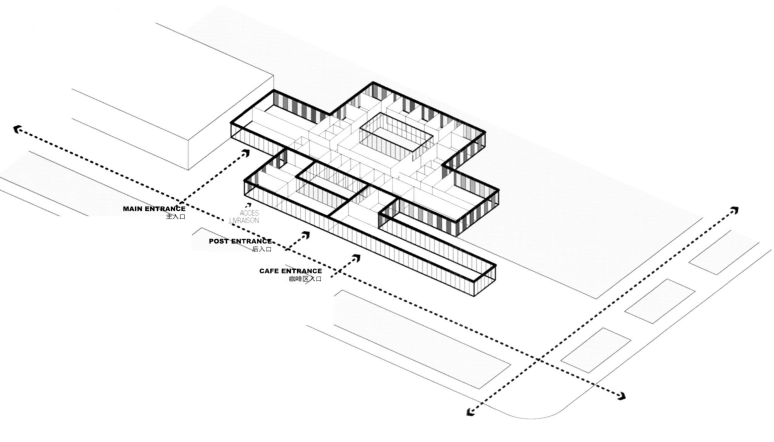

Rehabilitation - Extension of the Library
整修—图书馆扩建

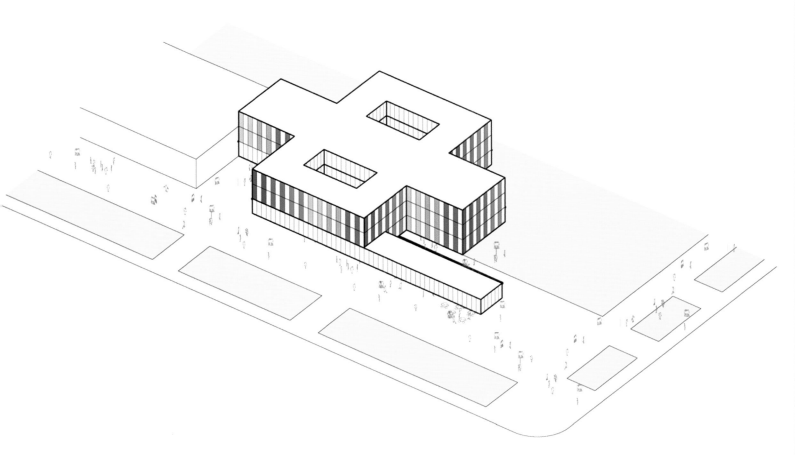

Rehabilitation - Extension of the Library
整修—图书馆扩建

Render - Aerial View
效果图—鸟瞰图

Render - New Entrance of the Mechanics Hall (ME Building)
效果图—机械学大厅新入口

Render - The Mechanics Hall (ME Building)
效果图—机械学大厅

Render - The Teaching Bridge and the Former Library (BI Building)
效果图—教学桥和原图书馆（BI 大楼）

Render - The Teaching Bridge
效果图—教学桥

Render - The Teaching Bridge from the Avenue Piccard
效果图—教学桥连接皮卡尔大道

福建厦门光电职业技术学院

The Fujian Professional Photonic Technical College

设计单位：10 DESIGN（拾稼设计）	Designed by: 10 DESIGN
开发商：庆富集团	Client: Ching Fu Group
项目地址：中国福建省厦门市	Location: Xiamen, Fujian, China
占地面积：1 050 227 ㎡	Site Area: 1,050,227 m²
建筑面积：524 000 ㎡	Gross Floor Area: 524,000 m²
建筑团队：Ted Givens Mohamad Ghamlouch Sonja Stoffels Ray Lam Tatsuya Sakairi Nkiru Mokwe Li Xi Bryan Diehl Lynn Kim Emre Icdem	Architecture Team: Ted Givens, Mohamad Ghamlouch, Sonja Stoffels, Ray Lam, Tatsuya Sakairi, Nkiru Mokwe, Li Xi, Bryan Diehl, Lynn Kim, Emre Icdem
景观团队：Ewa Koter Ibrahim Diaz Shingrong Wu	Landscape Team: Ewa Koter, Ibrahim Diaz, Shingrong Wu

**Geometric Design,
Symbiosis Environment,
Green Ecology**

**几何设计
环境共生
绿色生态**

设计理念

项目的核心设计理念是将这所大规模的学院融入周边的自然环境，同时保留并加强基地的现有特征。为此，设计师设计出一种非侵入式的建筑结构，尽可能地达到这一结构与学校高效的功能性之间的平衡，构建一个自然与人工创造平衡的未来可持续性校园。

设计特色

这个具有突破性意义的设计得益于项目极其特殊的地理位置。在设计过程中，设计师不是只遵循几何设计秩序，而是采取了顺应基地特征的工作方式，并据此制定了规划策略。为了尽可能地方便施工并增加规划的灵活性，设计师按常规方式将园区内的3条天然溪流改造为暗渠，并将其覆盖。绿色低碳、可持续发展技术以及被动式太阳能技术的运用，则进一步强化了规划的可行性。

绿色设计

项目的总体规划以被动式太阳能原理为依据，将一系列高科技创新的绿色低碳技术运用到配套设施中，并加以强化，使该项目成为一个系统的新型产品试验台。项目采用的高新绿色低碳设计包括光催化纳米涂料、海藻空气净化系统、有机肥料制造技术、热冷却系统、绿色墙体、水力和太阳能发电等。

主学术大楼将采用海藻空气净化系统，这一系统将从地下车库抽出的废气注入建筑幕墙上的海藻管中，与此同时，启动幕墙上的光催化涂料并计算出维持夜间空气清洁反应所需的紫外光照量，再通过太阳能电池在夜间发出紫外光，从而确保主学术大楼空气清洁反应可24小时持续进行。

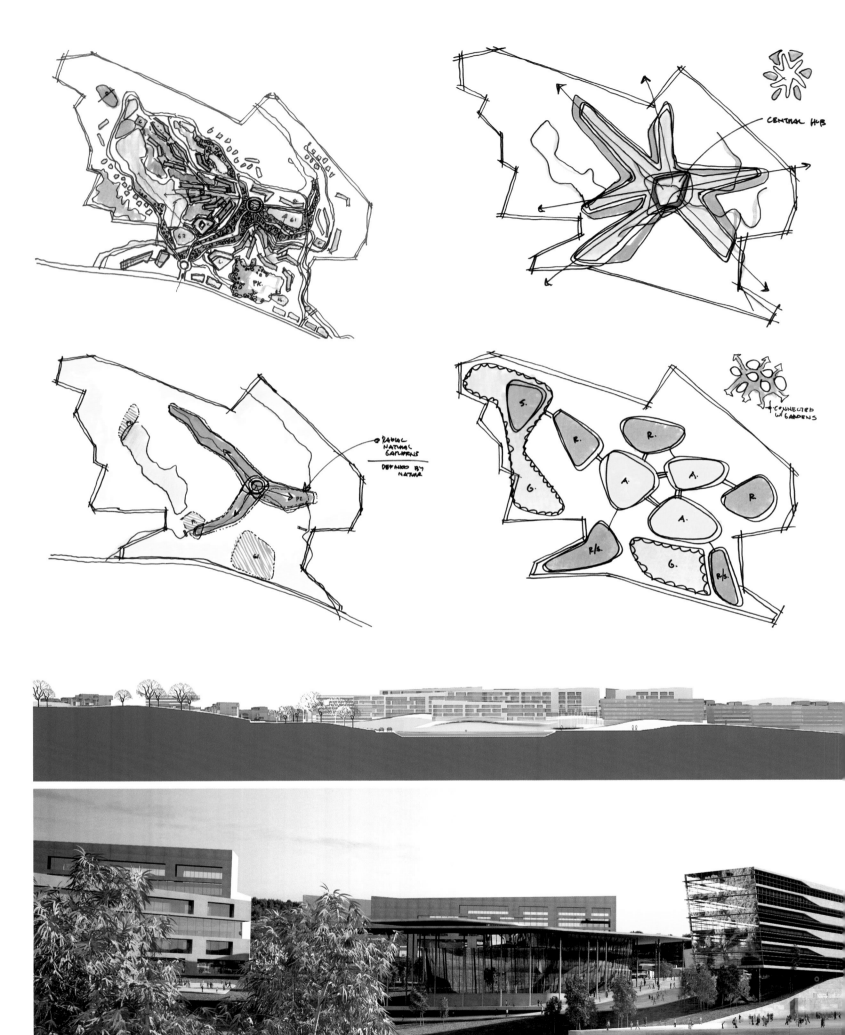

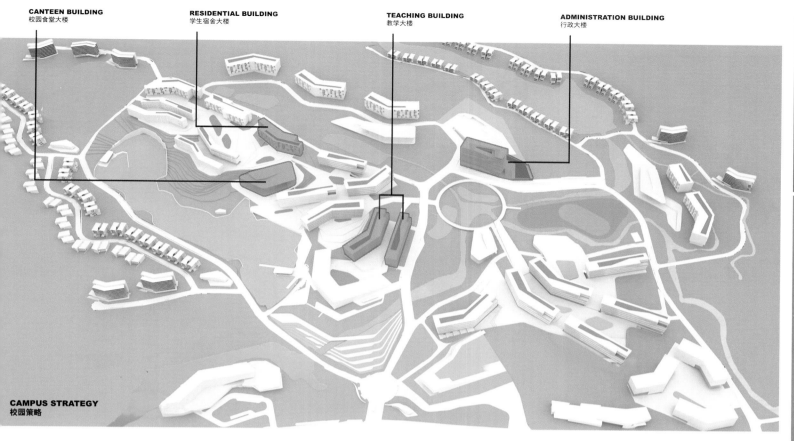

CANTEEN BUILDING
校园食堂大楼

RESIDENTIAL BUILDING
学生宿舍大楼

TEACHING BUILDING
教学大楼

ADMINISTRATION BUILDING
行政大楼

CAMPUS STRATEGY
校园策略

Design Concept

The core concept is to interweave a very large scale college into a natural setting, preserving and enhancing the sites existing character. Designers seek to design the most non-invasive set of structures possible, balanced against the creation of a very efficient college. The goal of the design is to plan the campus by putting nature and the creation of a balanced sustainable future first.

Design Feature

The significance of designing Fujian Professional Photonic Technical College benefits from its special geographic position. Instead of imposing a geometric order, designers worked with the site features and let them inform the campus planning strategy. The three streams would have normally been piped and covered to maximize ease of construction and planning flexibility. Low-carbon, sustainable development and passive solar energy technology in the design will further strengthen the project's feasibility.

Green Design

The master plan is built upon a backbone of passive solar principles that are supplemented and enhanced by a series of highly innovative sustainable technologies including the use of photocatalytic nano-coatings, algae for air purification and for the production of organic fertilizer, thermal cooling systems, green walls, hydro power, and solar power. The Photonic College will become test beds for the development of new products and systems.

The main academic building takes the algae air purification system in its design. The design will pump the exhaust from the underground parking up into a series of algae filled tubes on the façade. Designers are sourcing the photocatalytic coatings for the façades and calculating the amount of UV light needed to keep the air cleaning reaction happening at night as well, using a series of PV cells to power the UV lights. The goal is for the main academic building to clean the air 24 hours a day.

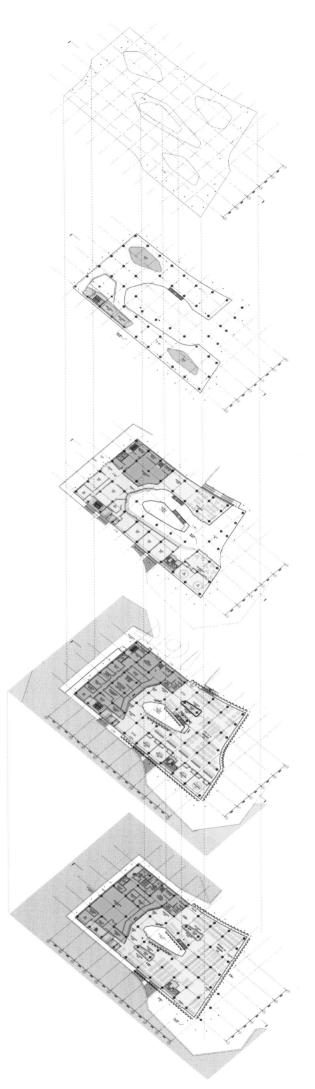

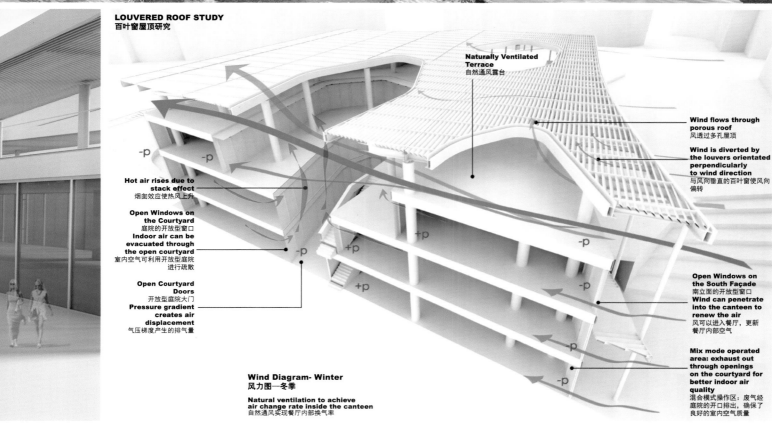

LOUVERED ROOF STUDY
百叶窗屋顶研究

Naturally Ventilated Terrace
自然通风露台

Wind flows through porous roof
风透过多孔屋顶

Wind is diverted by the louvers orientated perpendicularly to wind direction
与风向垂直的百叶窗使风向偏转

Hot air rises due to stack effect
烟囱效应使热风上升

Open Windows on the Courtyard
庭院的开放型窗口
Indoor air can be evacuated through the open courtyard
室内空气可利用开放型庭院进行疏散

Open Courtyard Doors
开放型庭院大门
Pressure gradient creates air displacement
气压梯度产生的排气量

Open Windows on the South Façade
南立面的开放型窗口
Wind can penetrate into the canteen to renew the air
风可以进入餐厅，更新餐厅内部空气

Mix mode operated area: exhaust out through openings on the courtyard for better indoor air quality
混合模式操作区：废气经庭院的开口排出，确保了良好的室内空气质量

Wind Diagram- Winter
风力图—冬季
Natural ventilation to achieve air change rate inside the canteen
自然通风实现餐厅内部换气率

办公建筑
Office Building

"炫酷灯笼"
流动立面
城市灯塔

"Cool Lantern",
Fluid Façade,
Urban Lighthouse

格鲁吉亚巴统法院

Palace of Justice of Batumi

设计单位：Architetto Michele De Lucchi S.r.l.	Designed by: Architetto Michele De Lucchi S.r.l.
开发单位：格鲁吉亚司法部	Client: Ministry of Justice of Georgia
项目地址：格鲁吉亚巴统	Location: Batumi, Georgia
总建筑面积：6 573 m²	Gross Floor Area: 6,573 m²
设计团队：Michele De Lucchi Alberto Bianchi Simona Agabio Marcello Biffi Greta Corbani Alessandra De Leonardis Laiza Tonali	Project Team: Michele De Lucchi, Alberto Bianchi, Simona Agabio, Marcello Biffi, Greta Corbani, Alessandra De Leonardis, Laiza Tonali
摄影：Gia Chkhatarashvili	Photography: Gia Chkhatarashvili

项目概况

巴统法院是该市城市开发中彰显城市特色的第一座大楼，包括部长办公室、公共注册中心、国家民事登记中心、巴统检察官和阿扎尔执行局等机关办事处。

设计特色

这是一座17层的标志性圆形塔楼，建筑造型独特，特别是构建在裙房上的塔楼，宛如悬挂在城市上空的灯笼，从周围的山峰或海岸上都可看到，这一建筑已成为一个具有导向和定位作用的"城市灯塔"。

建筑的外部饰面十分醒目，包括透明的观景模块和不透明模块，其中不透明的模块采用白色的阿鲁克邦复合材料面板，与白色的铝质材料一起突出了建筑的形态，无论白天还是夜晚，建筑发光的表面使建筑产生强烈的视觉冲击力。

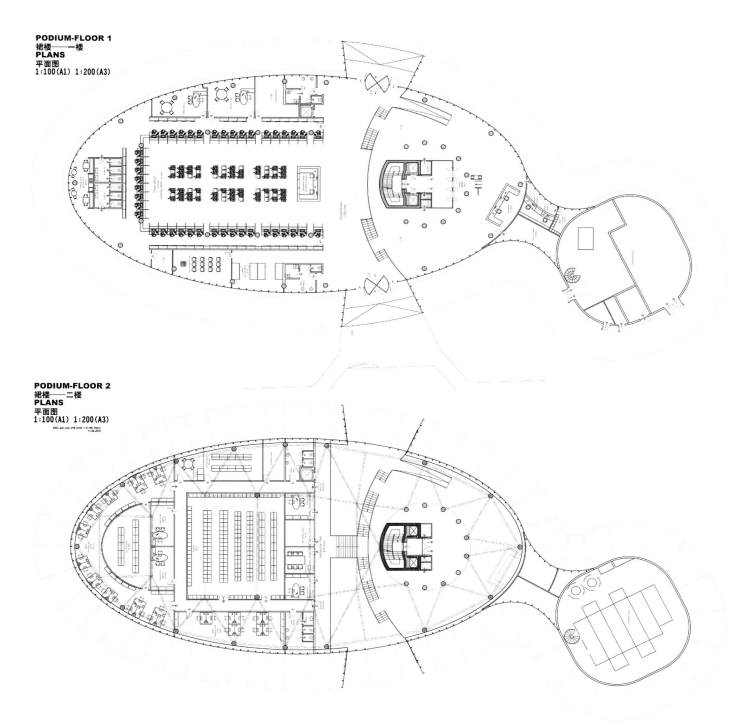

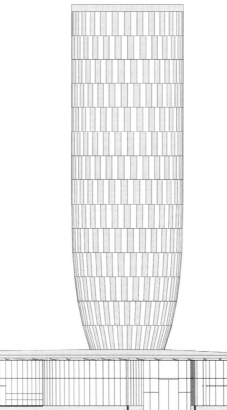

▽ +66.30 **Floor 17_Roof Terrace**
17 层屋顶露台

▽ +62.30 **Floor 16_Minister**
16 层部长办公区

▽ +58.30 **Floor 15_Prosecutor's Office of Autonomous Republic of Adjara**
15 层阿扎尔自治共和国检察官办公室

▽ +54.30 **Floor 14_Prosecutor's Office of Autonomous Republic of Adjara**
14 层阿扎尔自治共和国检察官办公室

▽ +50.30 **Floor 13_Batumi Regional Prosecutor's Office**
13 层巴统地区检察官办公室

▽ +46.30 **Floor 12_ Batumi Regional Prosecutor's Office**
12 层巴统地区检察官办公室

▽ +42.30 **Floor 11_Chamber of Financial Control**
11 层财务管理室

▽ +38.30 **Floor 10_Chamber of Financial Control**
10 层财务管理室

▽ +34.30 **Floor 9_Enforcement Bureau of Adjara**
9 层阿扎尔执行局

▽ +30.30 **Floor 8_ Enforcement Bureau of Adjara**
8 层阿扎尔执行局

▽ +26.30 **Floor 7_Legal Consultation Service**
7 层法律咨询室

▽ +22.30 **Floor 6_NACR**
6 层 NACR

▽ +18.30 **Floor 5_Adjara Justice Department**
5 层阿扎尔司法部

▽ +14.30 **Floor 4_Archive**
4 层档案室

▽ +10.30 **Floor 3_Server Room**
3 层服务器机房

▽ +5.50 **Floor 2_NAPR**
2 层 NAPR

▽ +1.00 **Floor 1_Reception and NACR**
1 层接待处和 NACR

Façade West
西立面

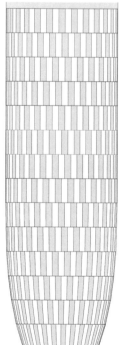

▽ +66.30 **Floor 17_Roof Terrace**
17 层屋顶露台

▽ +62.30 **Floor 16_Minister**
16 层部长办公区

▽ +58.30 **Floor 15_Prosecutor's Office of Autonomous Republic of Adjara**
15 层阿扎尔自治共和国检察官办公室

▽ +54.30 **Floor 14_Prosecutor's Office of Autonomous Republic of Adjara**
14 层阿扎尔自治共和国检察官办公室

▽ +50.30 **Floor 13_Batumi Regional Prosecutor's Office**
13 层巴统地区检察官办公室

▽ +46.30 **Floor 12_Batumi Regional Prosecutor's Office**
12 层巴统地区检察官办公室

▽ +42.30 **Floor 11_Chamber of Financial Control**
11 层财务管理室

▽ +38.30 **Floor 10_Chamber of Financial Control**
10 层财务管理室

▽ +34.30 **Floor 9_Enforcement Bureau of Adjara**
9 层阿扎尔执行局

▽ +30.30 **Floor 8_Enforcement Bureau of Adjara**
8 层阿扎尔执行局

▽ +26.30 **Floor 7_Legal Consultation Service**
7 层法律咨询室

▽ +22.30 **Floor 6_NACR**
6 层 NACR

▽ +18.30 **Floor 5_Adjara Justice Department**
5 层阿扎尔司法部

▽ +14.30 **Floor 4_Archive**
4 层档案室

▽ +10.30 **Floor 3_Server Room**
3 层服务器机房

▽ +5.50 **Floor 2_NAPR**
2 层 NAPR

▽ +1.00 **Floor 1_Reception and NACR**
1 层接待处和 NACR

Façade East
东立面

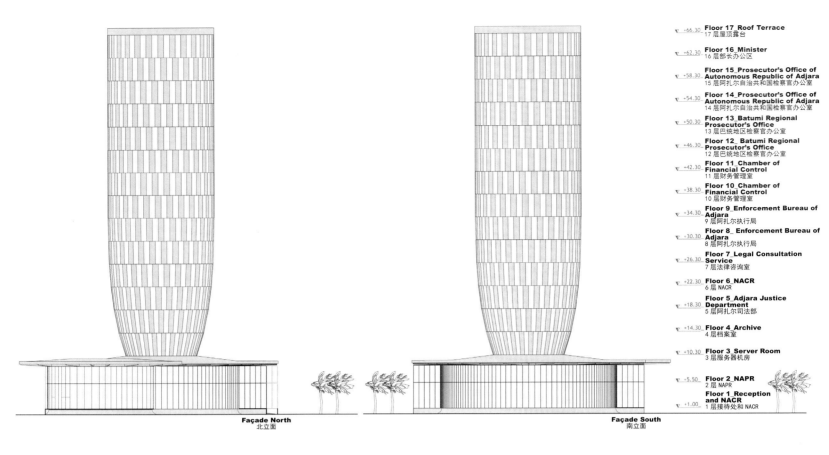

Elev.	Floor	Description
+66.30	Floor 17	Roof Terrace / 17层屋顶露台
+62.30	Floor 16	Minister / 16层部长办公区
+58.30	Floor 15	Prosecutor's Office of Autonomous Republic of Adjara / 15层阿扎尔自治共和国检察官办公室
+54.30	Floor 14	Prosecutor's Office of Autonomous Republic of Adjara / 14层阿扎尔自治共和国检察官办公室
+50.30	Floor 13	Batumi Regional Prosecutor's Office / 13层巴统地区检察官办公室
+46.30	Floor 12	Batumi Regional Prosecutor's Office / 12层巴统地区检察官办公室
+42.30	Floor 11	Chamber of Financial Control / 11层财务管理室
+38.30	Floor 10	Chamber of Financial Control / 10层财务管理室
+34.30	Floor 9	Enforcement Bureau of Adjara / 9层阿扎尔执行局
+30.30	Floor 8	Enforcement Bureau of Adjara / 8层阿扎尔执行局
+26.30	Floor 7	Legal Consultation Service / 7层法律咨询室
+22.30	Floor 6	NACR / 6层 NACR
+18.30	Floor 5	Adjara Justice Department / 5层阿扎尔司法部
+14.30	Floor 4	Archive / 4层档案室
+10.30	Floor 3	Server Room / 3层服务器机房
+5.50	Floor 2	NAPR / 2层 NAPR
+1.00	Floor 1	Reception and NACR / 1层接待处和NACR

Façade North 北立面

Façade South 南立面

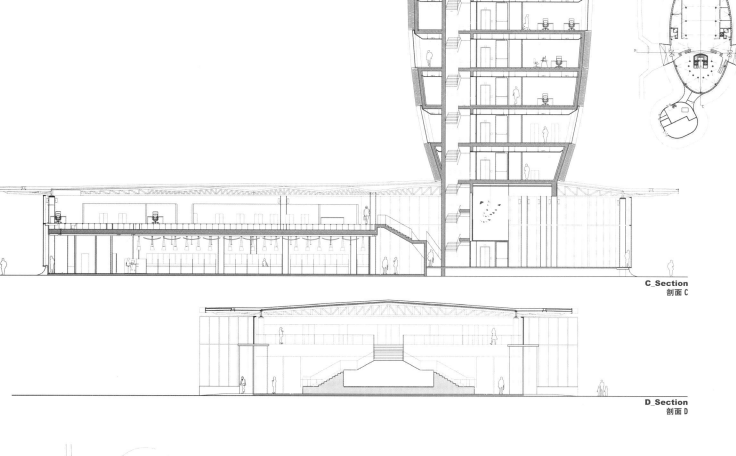

C_Section
剖面 C

D_Section
剖面 D

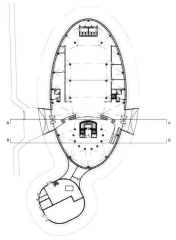

Profile

The Palace of Justice in Batumi is the first building of the urban development that characterizes an important area of the city. The public palace will include the offices of the Minister, the Public Registry Agency, the National Civil Registry Agency, the Procurator of Batumi and the Enforcement Bureau of Adjara.

Design Feature

The 17-level symbolic round tower has a unique shape. The tower above the podium resembles a lantern hanging above the city. The new building will be visible from the hills and the sea. It will be a landmark for the people, a typical lighthouse that doesn't bother the landscape but oriented and qualifies itself.

The tower has remarkable external finish consisting of vision and opaque modules with the latter made from white Alucobond panels. The white Alucobond panels and the aluminum profiles further emphasize the building's harmonious and innovative shape. It has strong visual impact both by day and at night due to its shiny surface.

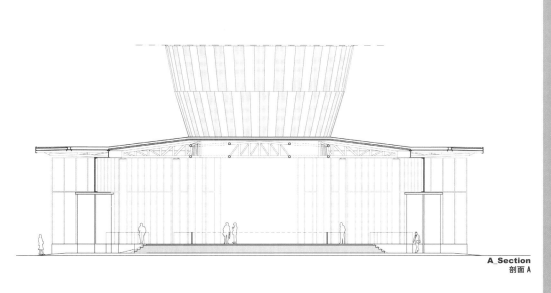

A_Section
剖面 A

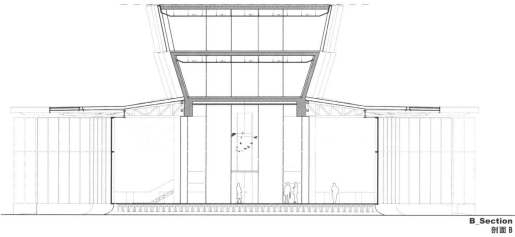

B_Section
剖面 B

日本东京比利时驻日大使馆改建
Belgian Embassy Reconstruction

设计单位：Philippe SAMYN and PARTNERS, Architects & Engineers
　　　　　KAJIMA DESIGN TOKYO
开发商：鹿岛建设
　　　　三井不动产
项目地址：日本东京
总建筑面积：39 955 m²（其中，使馆大楼 6 227 m²，办公楼 33 728 m²）
平面图和手绘图：Philippe SAMYN and PARTNERS, Architects & Engineers
　　　　　　　　KAJIMA DESIGN TOKYO

Designed by: Philippe SAMYN and PARTNERS, Architects & Engineers; KAJIMA DESIGN TOKYO
Client: Kajima; Mitsui Fudosan
Location: Tokyo, Japan
Gross Floor Area: 39,955 m²(Embassy Building: 6,227 m²; Office Building: 33,728 m²)
Plans and Drawings: Philippe SAMYN and PARTNERS, Architects & Engineers; KAJIMA DESIGN TOKYO

**空中花园
生态环保
空间功能
最大化**

**Air Garden,
Eco-Environment
Protection,
Maximized Space
Function**

项目概况

比利时驻日大使馆位于东京 Bancho 地区，临近皇宫。整个项目包括新的大使馆大楼以及与大使馆位于同一场址、与环境共生的新办公大楼。方案中广泛运用的绿色设计使项目达到了日本建筑物综合环境性能评价"S"等级的标准。

设计构思

要在日本东京这样一个高度集中且面积相对狭小的大都市里营造一个宁静、绿色的环境实属不易，却也不是不可能，本方案就为实现这一目标提供了一个参照。大使馆周围绿树成荫，故设计师将项目定义为一个新的广场、一个大型的花园，同时也是一座绿色建筑，这一设想充分利用了场地优越的条件，并以积极的态度、富有影响力的方式投身到保护城市环境的大业之中，完成了对当今城市环境的阐释以及对该地区未来的展望。

设计主要基于建筑的身份象征以及建筑的自主性、可持续设计、环保性、私密性和安全性来展开，从而构建一个具有场所感和时尚感，自然生态、可持续的平衡性建筑。

设计特色

设计师将最大限度地延展空间的功能，实现对空间的最大化利用。建筑入口可以改造成为剧场，楼梯也可作为放映通道，这些空间在功能上的延展，既体现了设计师对空间的灵活把握，同时，因这些空间皆对当地社区开放，也提高了空间的可参与性。

可持续性设计

住宅楼层设置有可从落地窗通向装有百叶窗的阳台，住户可通过调整百叶窗来控制采光，从而营造适合个人尺度的、舒适的室内环境。大使馆大楼的外观上也设计了可调节的水平百叶窗，在实现自然通风的同时，不会干扰到建筑的景观视野。另外，花园露台和低辐射双层玻璃也有效地减少了建筑的热负荷。

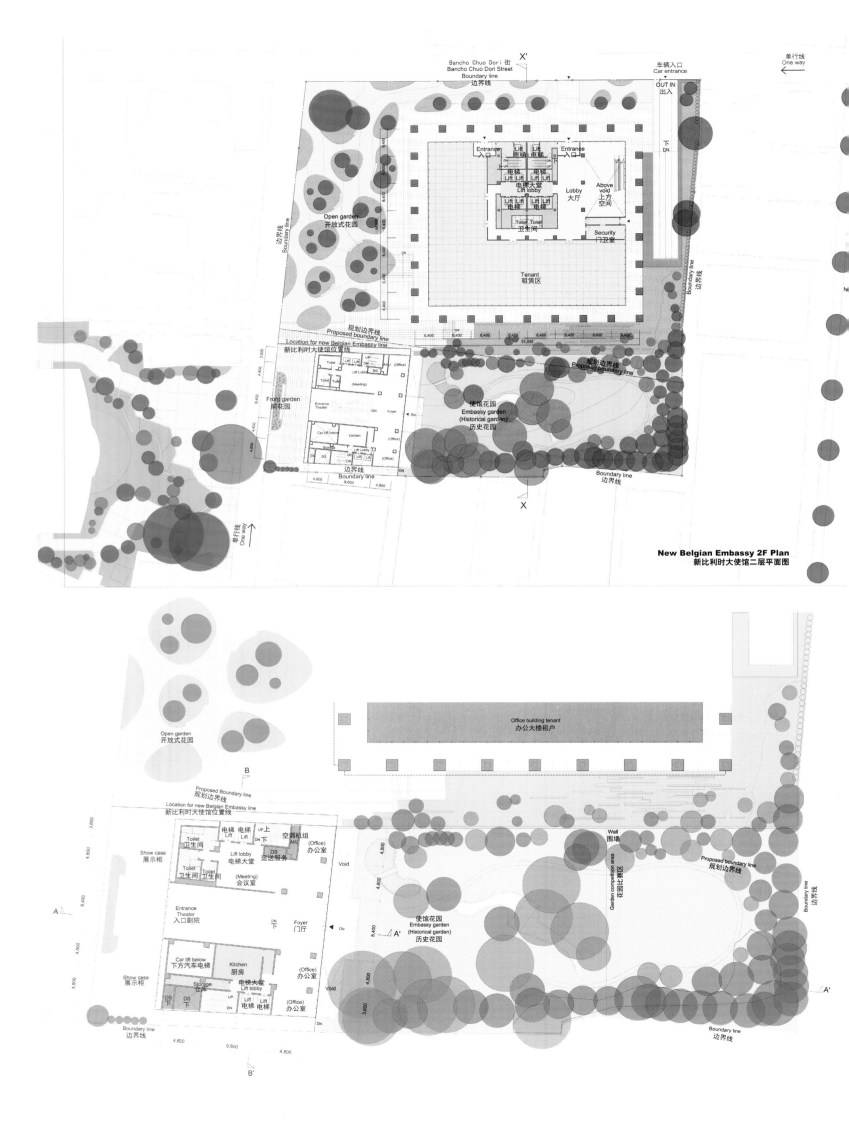

B2F Plan
地下室二层平面图

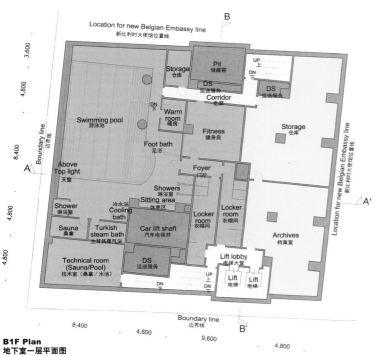

B1F Plan
地下室一层平面图

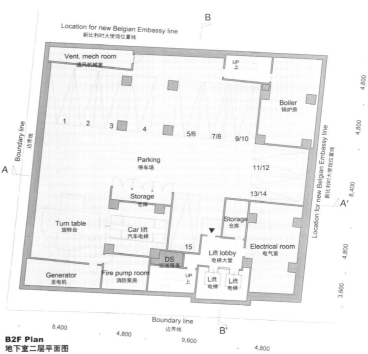

B2F Plan
地下室二层平面图

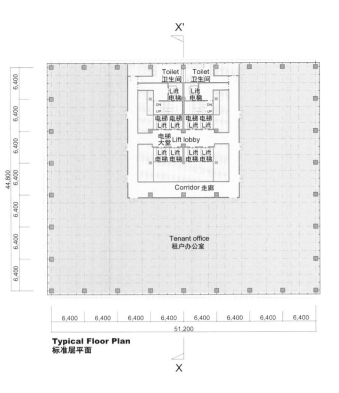

Typical Floor Plan
标准层平面

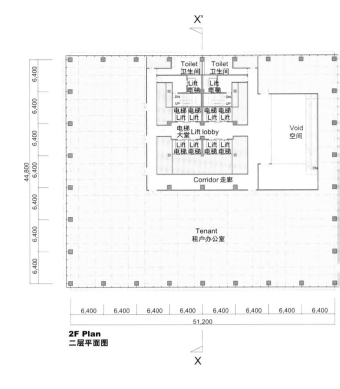

2F Plan
二层平面图

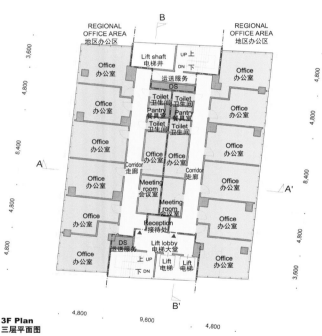

3F Plan
三层平面图

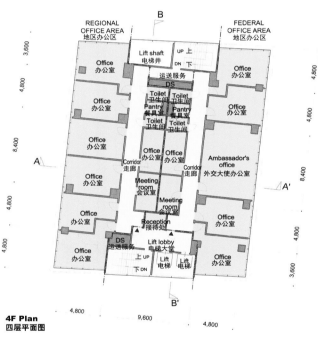

4F Plan
四层平面图

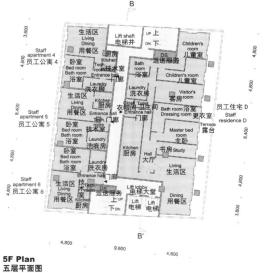

5F Plan
五层平面图

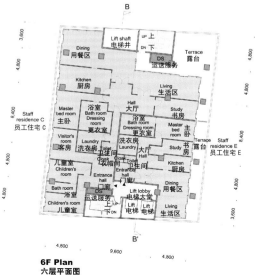

6F Plan
六层平面图

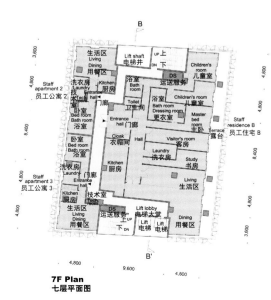

7F Plan
七层平面图

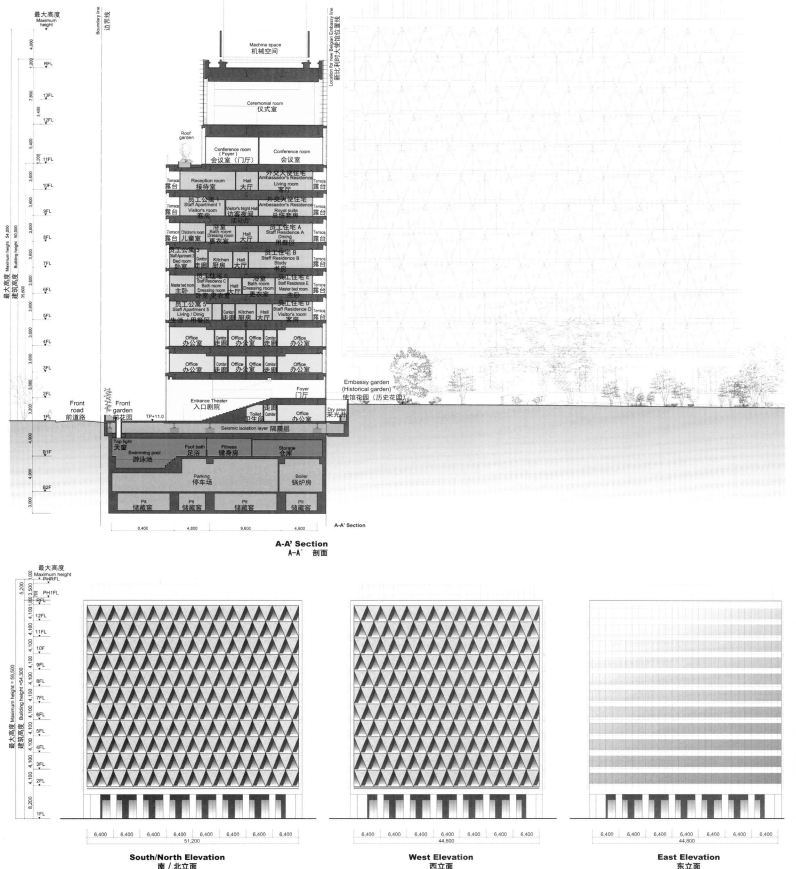

Profile

The project is located in the Bancho district in the vicinity of Imperial Palace. It includes the new Embassy building and offers the design concept of a new office building that coexists harmoniously on the same site as the embassy. The design scheme for this project achieved Rank "S" by the evaluation of the Japanese "Comprehensive Assessment System for Building Environmental Efficiency" (CASBEE).

Design Concept

To create a tranquil green environment in a highly concentrated urbanized area of Tokyo with limited land is rare but possible. This scheme provides a reference to achieve such ambition. Blessed with green vegetations, the project defines a new plaza, a new vast garden and also a literally green building. The use of the site is arranged to contribute, actively and in an attractive way, to the urban environment and the local community. Based on a careful interpretation of the social and urban context of the site and the future prospects of the area, the project implements a balanced architectural approach.

The design, based on the Embassy's clear identity and architectural autonomy, sustainability, environmental consciousness, privacy and security, plans to construct a sustainable balanced building with concepts of "Place", "Nature", "Sense of Beauty" and "Time".

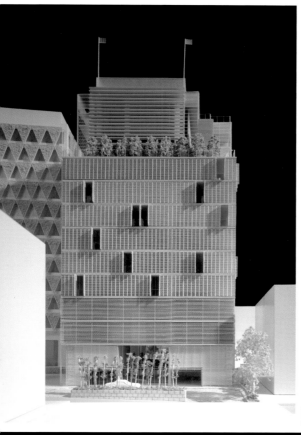

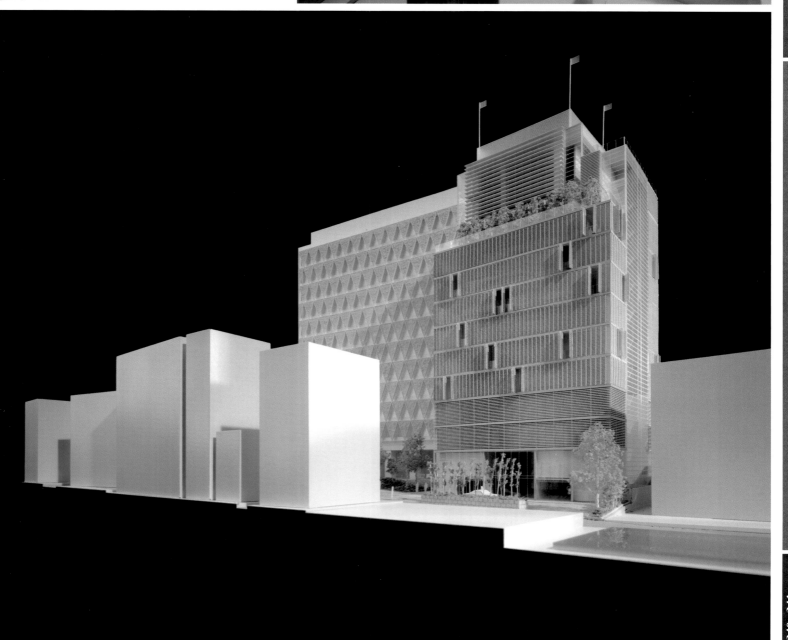

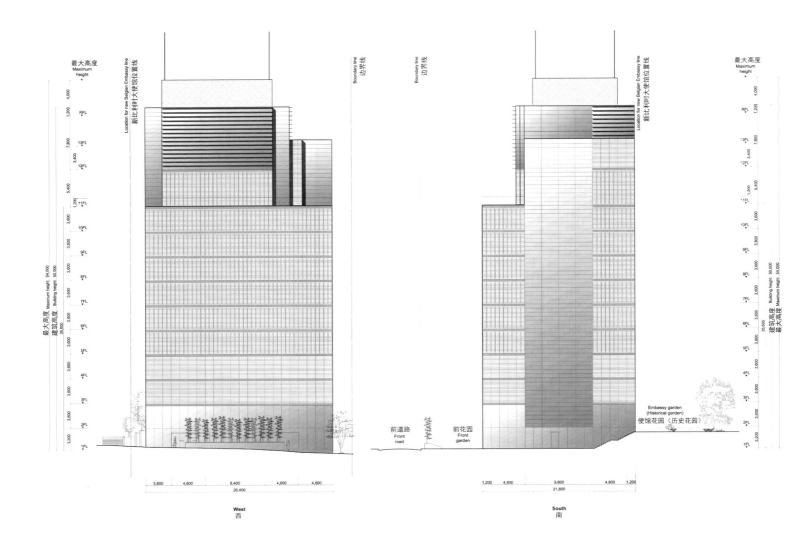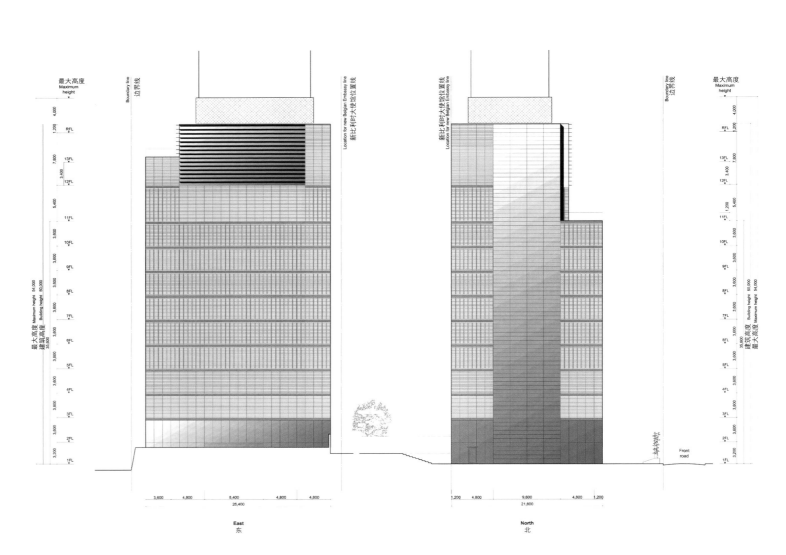

Design Feature
Designers maximally extend space functions to achieve maximum utilization of spaces. The entrance of the Embassy building can be transformed into a theatrical space using stairs as aisles for a projection. Public spaces are open to the local community, where various events can take place. This way, the design achieves the extension of functions and flexible use of spaces.

Sustainable Design
On residential floors, louvered balconies are accessible through French windows. Occupants can adjust daylight through louvered doors to obtain comfortable interior environment. Adjustable horizontal louvres are also installed on the façade of the Embassy building to achieve natural ventilation and unobstructed views. In addition, gardened terraces and low-e double glazing effectively reduce thermal loads.

葡萄牙 Maia LÚCIO 新总部办公大楼

设计单位：João Álvaro Rocha – Arquitectos
项目地址：葡萄牙 Maia
总建筑面积：6 341 m²
摄影：Luís Ferreira Alves

Designed by: João Álvaro Rocha – Arquitectos
Location: Maia, Portugal
Gross Floor Area: 6,341 m²
Photography: Luís Ferreira Alves

设计构思

这里有着喧闹的街道，具有一定规模和场所感的购物中心、可用面积很小的地块，粗略看去，整个场地似乎是众多杂乱而无序的事物的组合。多样化的建筑体零散分布，不能形成一个连续的序列，这也使得集体住宅建筑、服务区和单个家庭住宅难以共存。那么，在这个"杂乱得无处容身"的地块里，要"插入"一栋办公大楼，就需要转换一下思维，以自我防御的姿态融入其中。

设计特色

建筑"多变"的形态源自合理的平面布局，这也避免了西侧繁华嘈杂的街道对建筑产生干扰。这个结构紧凑的弧形体量宛如一个有着些微开口的"蚌壳"，有着连续的表面，虽缺失了所谓的"正立面"或任何重要的"侧面"，却为建筑在这个极度"排外"的场地里赢得了一块"立身之处"。

建筑表面的开口打破了整体的连续性，这一开口从外立面处呈一定的角度向建筑内部延伸，形成一个开放的公共空间，类似一个小型的广场，既为职工提供了一个休闲的场所，也可将更多的自然光线引入建筑内部。这一独特的开口定义了建筑的入口空间，也丰富了建筑的表现形式。

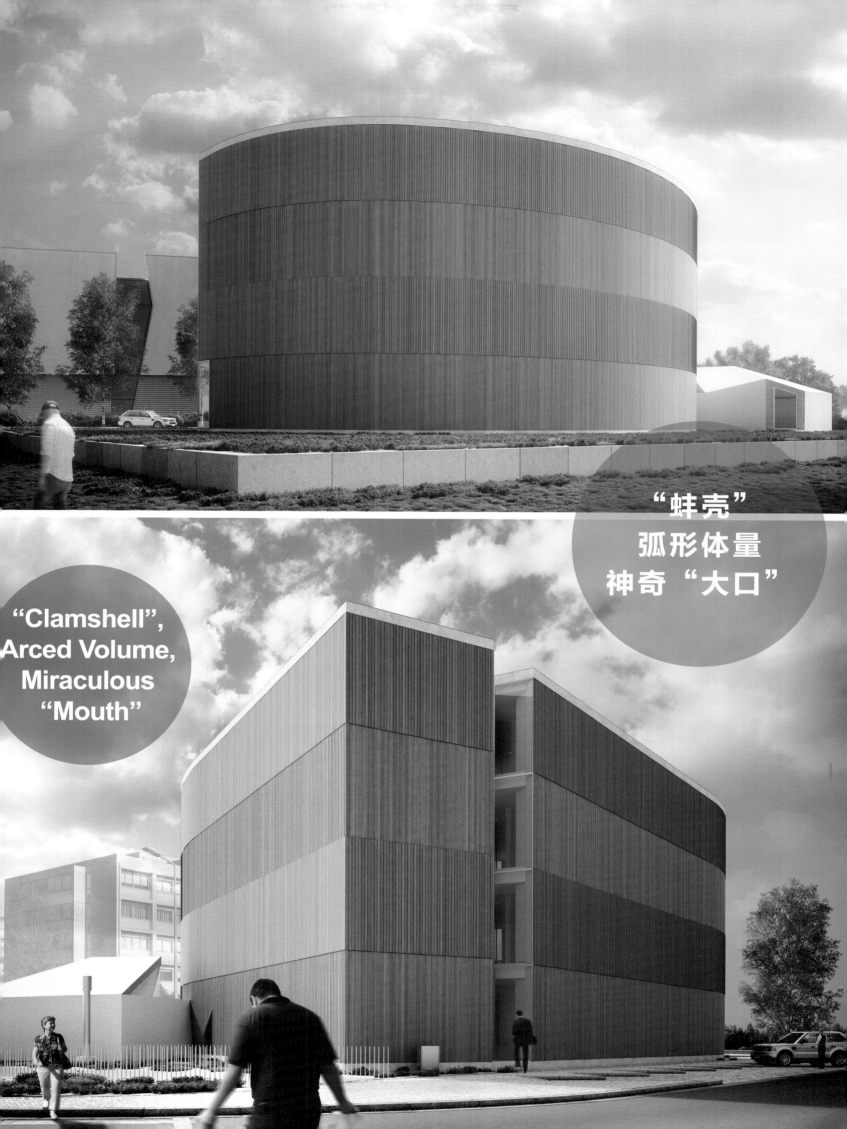

"蚌壳"
弧形体量
神奇"大口"

"Clamshell",
Arced Volume,
Miraculous
"Mouth"

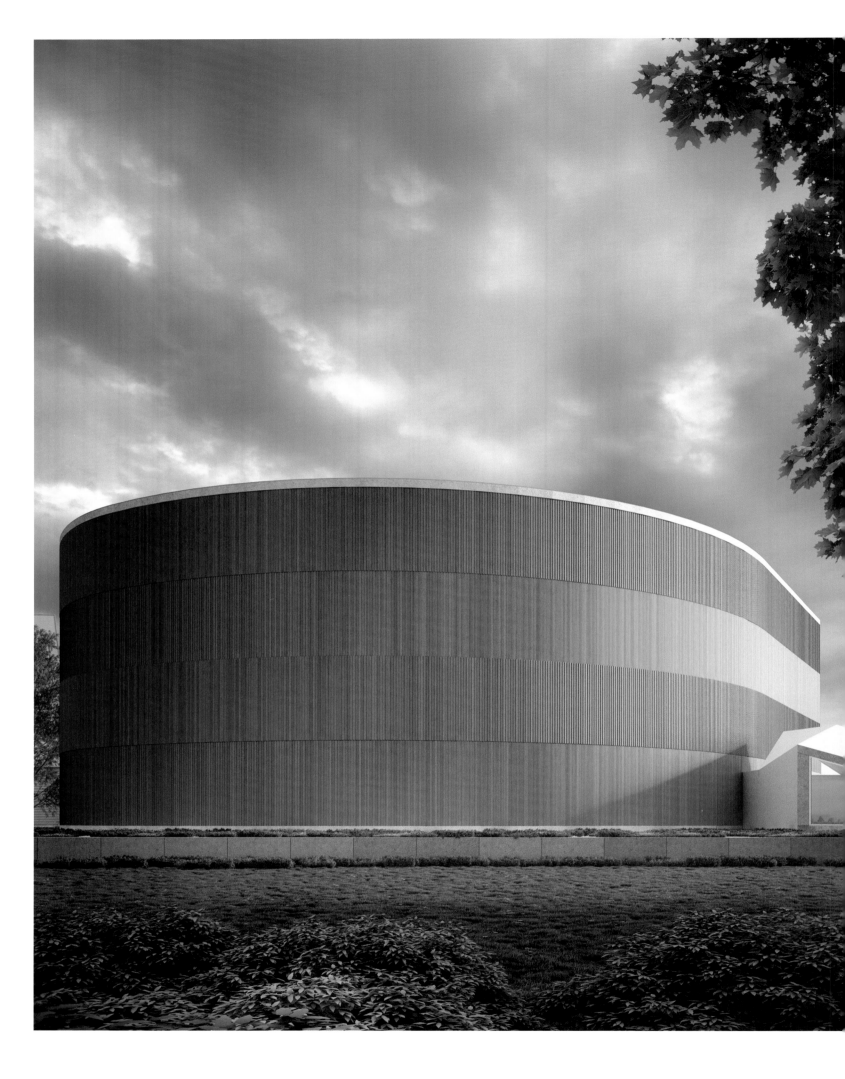

ARRUAMENTO (BETUMINOSO)   CUBO DE GRANITO 11x11   CUBO DE GRANITO 5x5   ÁREAS AJARDINADAS   EDIFÍCIOS   GRAVILHA

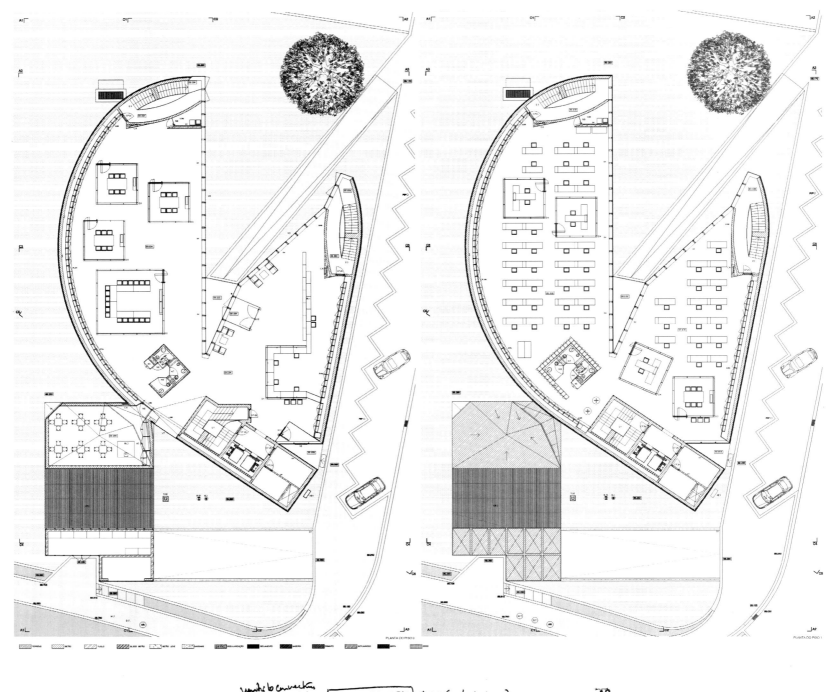
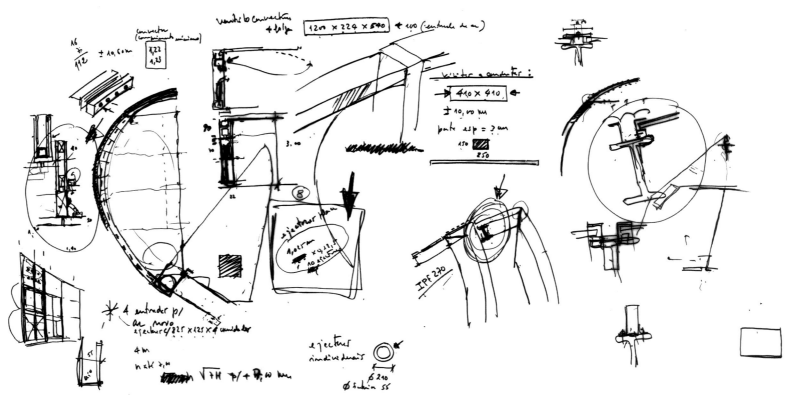

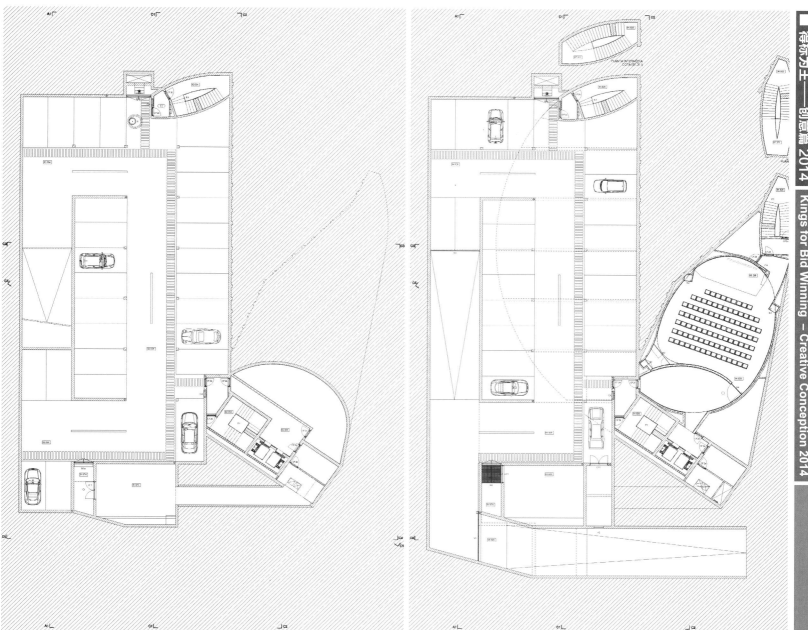
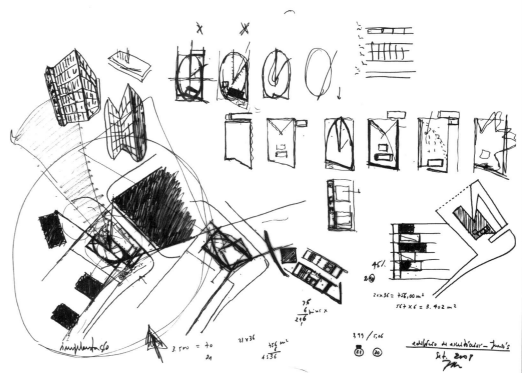

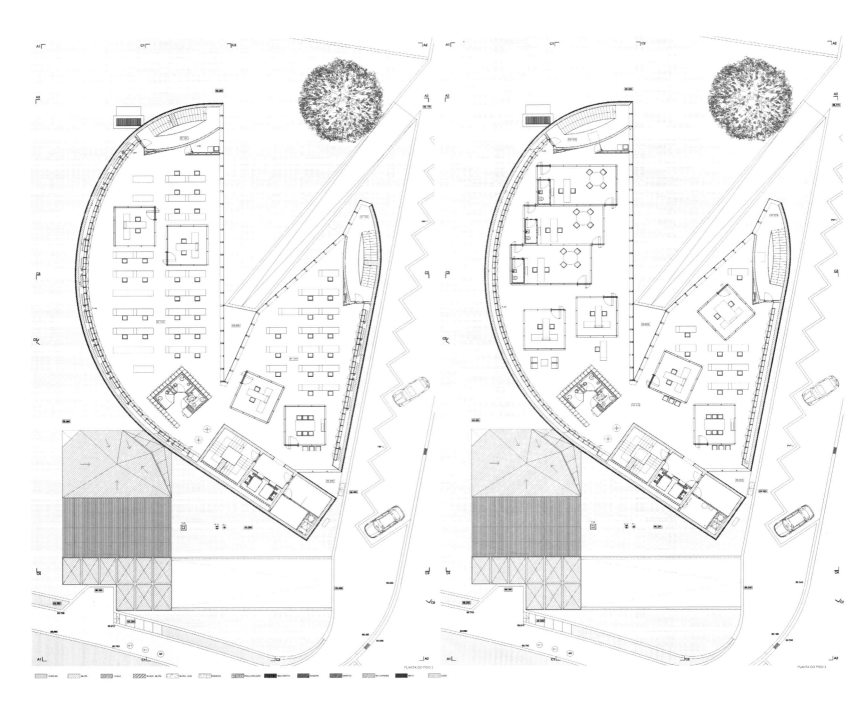
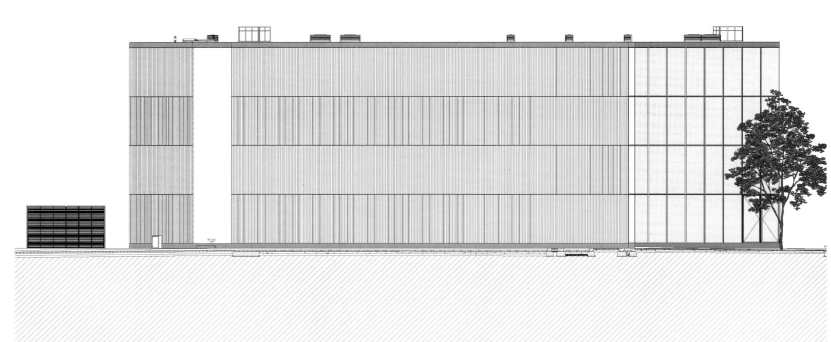

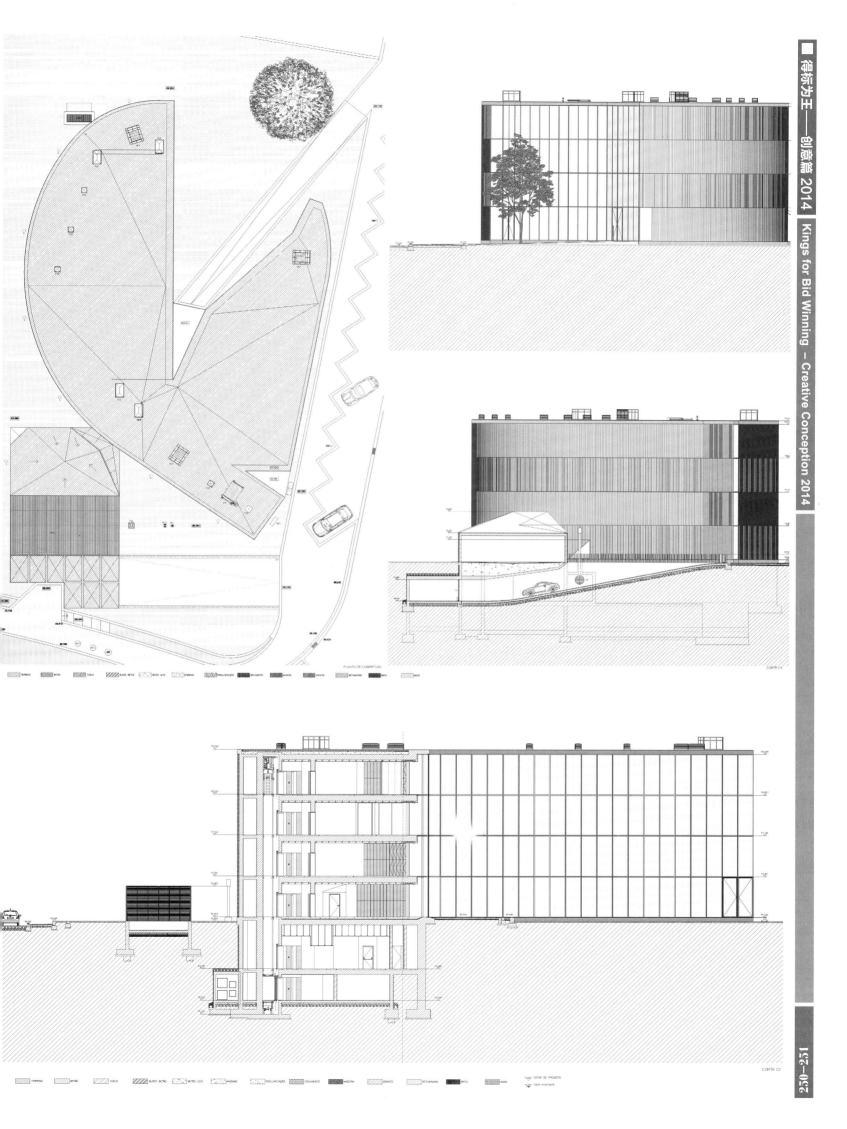

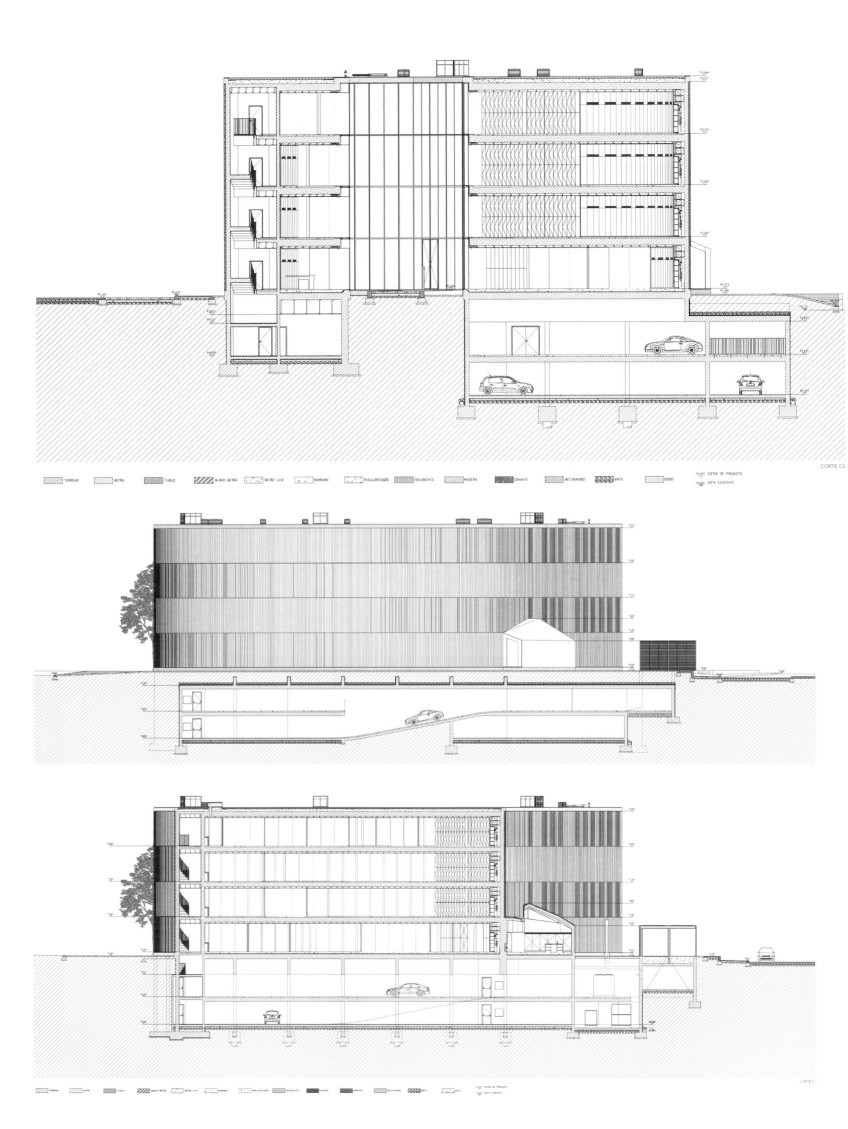

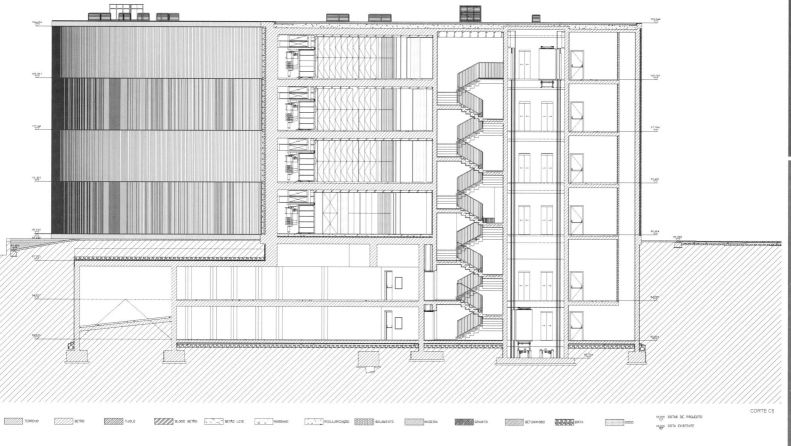

Design Concept

Noisy streets, shopping center of strong presence and certain scale, and small portion of available plot, the whole site seems to be a combination of disordered stuffs. A scattered layout of various volumes is unable to form a continuous sequence whilst makes collective residential buildings, services, single family homes difficult to coexist. Therefore, to "insert" an office building in this messy context (seems to have no "place"), the project must necessarily adopt a defensive attitude.

Design Feature

The "capricious" form of the building results from the possible alignments and from the desire to protect it from the busy and noisy road which lies to the west. Its compact arced volume looks like a "clamshell" with continuous surfaces slightly opening its mouth. In a way that eliminate the existence of any "front" or any important "side", the project gains the autonomy it needs to "survive" with dignity in such an "antagonistic" context.

The opening on the façade of the building breaks its continuity. It extends to the internal of the building with certain angle to create an open public space, like a small square, which can provides a leisure place for staffs and introduces daylight to the internal. The opening defines the main entrance and enriches expression of the building.

伊朗布什尔建筑工程组织大楼
Bushehr Construction Engineering Organization Building

设计单位：Process-based Architecture Studio	Designed by: Process-based Architecture Studio
开发单位：布什尔建筑工程组织	Client: Bushehr Construction Engineering Organization
项目地址：伊朗布什尔	Location: Bushehr, Iran
设计团队：Jafar Bazzaz,	Design Team: Jafar Bazzaz,
Arash Pouresmaeil,	Arash Pouresmaeil,
Kamal Youssefpoor	Kamal Youssefpoor

Wooden Skin, Urban "Jungle", Mobile Façade

木质表皮
城市"丛林"
活动立面

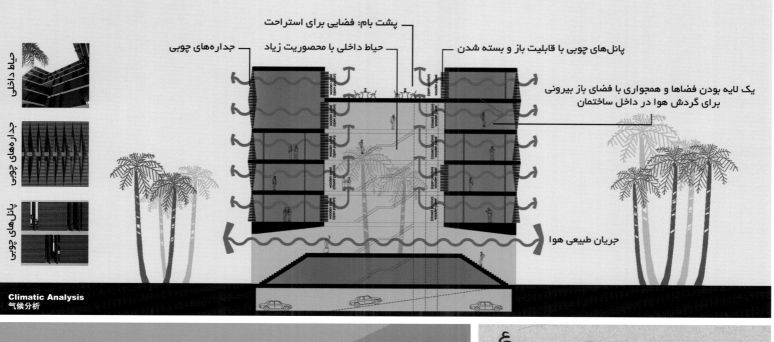

Climatic Analysis
气候分析

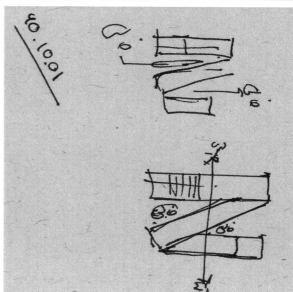

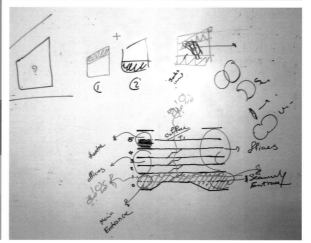

设计构思

方案的设计以场地内两个基本的条件为基础：内部空间和外部开放空间之间的过渡区域的对流通风是针对当地温暖湿润的气候而设计的，而建筑的结构和布局则与通过场地的两条主要道路和次要道路息息相关。

建筑设计

整个建筑由两个分离的部分组成，分别位于场地交通线路的两侧，使项目具有良好的交通可达性。两个体量中间有一个高度较低的连接结构，这样的布局结构形成了两个小型的内部庭院，而周围高高的围墙以及木质立面，使庭院成为有荫蔽的、宜人的开放空间，恰似一座迷人的城市丛林。

在温暖潮湿的气候条件下，采用低热容量的材料是十分必要的。因此，该项目的外墙采用了两种类型的木质表皮，其中面向街道的外墙木质表皮有着独特的结构和几何形态，而面向内部庭院的外墙木质表皮可根据需求打开或关闭。

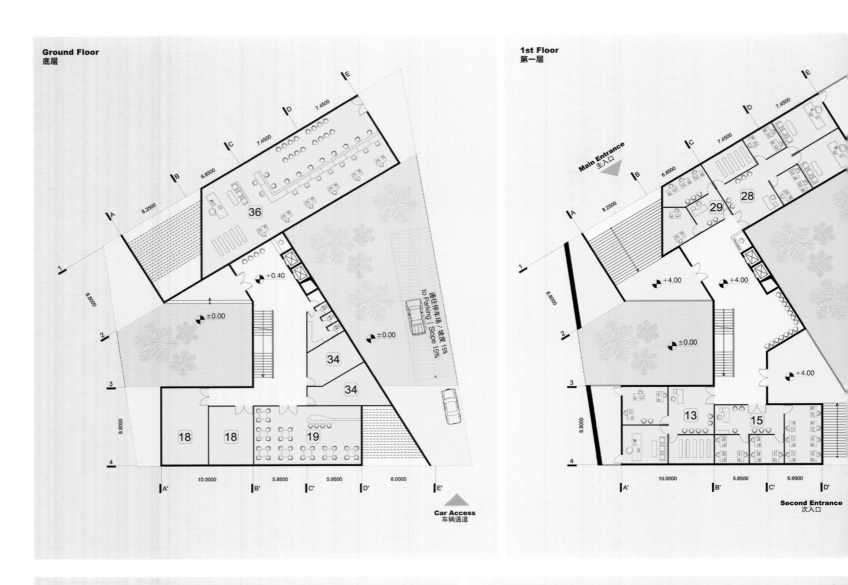
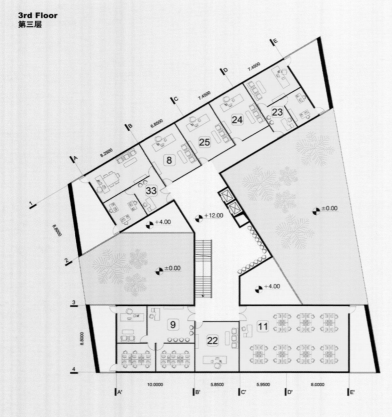
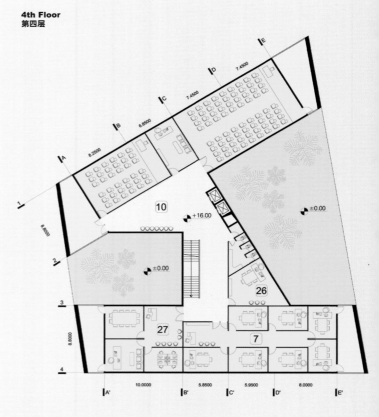

Design Concept

The proposal is based on design constraints of warm and humid climate and specific accessibility of the project's site through two major and minor roads. Cross ventilation through the spaces adjacent to internal and external open spaces is the most fundamental climatic design principle of this building.

Architectural Design

The building is organized by separated parts in two accessible sides of the site that are connected through a middle part with lower height. This organization has been formed two internal small yards with high surrounding walls that create pleasant shaded open spaces.

Due to the necessity of using materials with low heat capacity in the warm and humid climate, two types of wooden skins are used in the exterior walls. In the walls facing the street, a wooden skin with special structure and geometry and in the walls facing the internal yards, wooden panels are used that could be opened and closed.

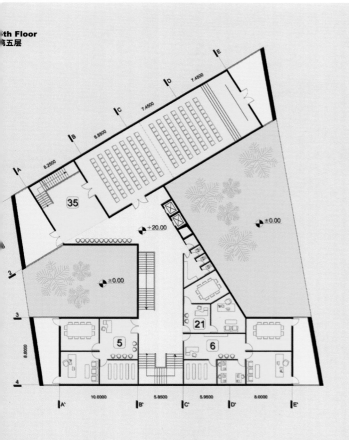

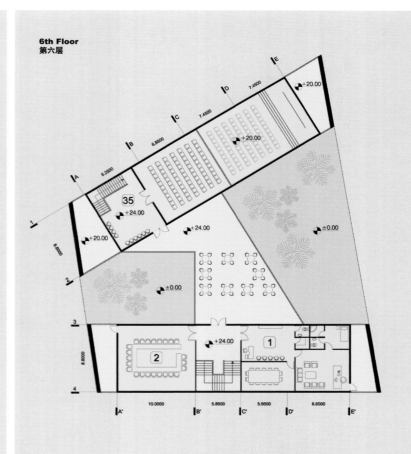

Design Problem?
设计问题?

First Side of Building
建筑第一面

Second Side of Building
建筑第二面

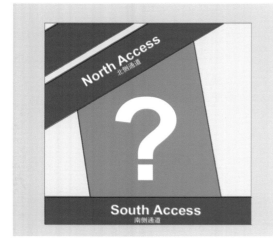
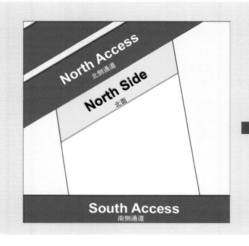
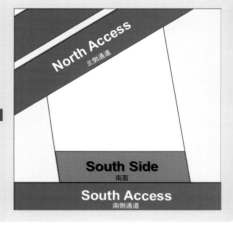

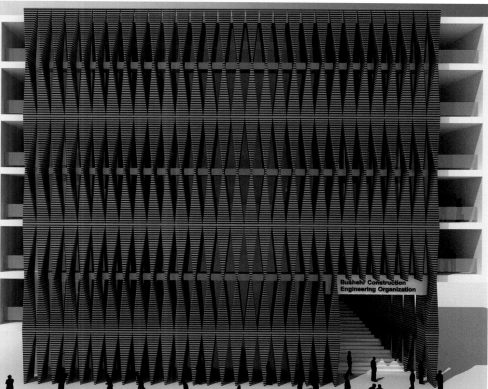

Two Sides of Building
建筑的两面

Integration of Two Sides
两面的整合

Integrated Sides
经整合的两面

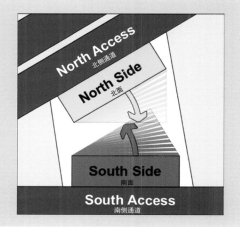
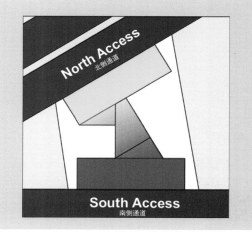

Concave & Convex, Double-skin Façade, Bright Gold

凹凸有致 双表皮立面 璀璨黄金

西班牙安达卢西亚自治区政府办公楼
Office Tower for the Government of Andalusia

设计单位：Guillermo Vazquez Consuegra Arquitecto S.L.P
开发商：Junta de Andalucía. Consejería de Economía y Hacienda.
项目地址：西班牙安达卢西亚自治区
建筑面积：41 797 ㎡

Designed by: Guillermo Vazquez Consuegra Arquitecto S.L.P
Client: Junta de Andalucía. Consejería de Economía y Hacienda.
Location: Andalusia, Spain
Floor Area: 41,797 m²

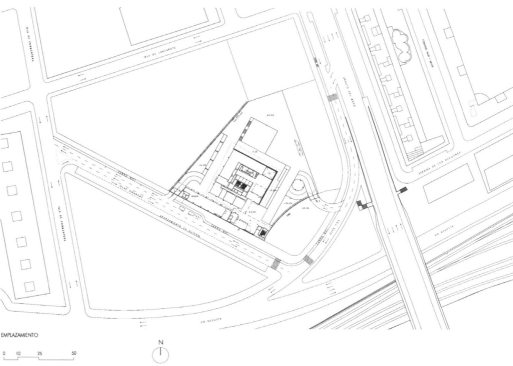

设计构思

项目场地位于科尔多瓦的郊区，处于这座现代化城市的新主干线上。原有的建筑毫无特色，且不具备独特的身份象征，故设计旨在为安达卢西亚地区政府构建一座独立而又精密的建筑，使之成为这一地区的新地标。

办公楼包括5个省级部门办事处以及各种多功能用途的公共区和停车场，考虑到行政办公楼功能的复合性，设计试图在赋予每一功能区独有特色的基础上，将之整合成一个整体，使其整体功能大于各部分的总和。

设计特色

省级办事处将占据3个连续的楼层，每个办事处的较低层都附有一个大型的门厅，门厅高3层，向位于该层角落处的花园开敞，是一个可进行交流、休憩的场所。门厅和花园之间是一部轻金属材质的楼梯，楼梯的设置不仅有利于内部交流，而且确保了每个办事处与花园之间的视觉联系。

模块化的办公空间可实现多元化的空间配置，保证内部空间的灵活性和多功能性，为此，设计师设想了一个中性的、统一的建筑外观，确保空间的多样化的配置。建筑的外立面上将覆盖一个玻璃外表皮，其外是金属晶格外表皮，双表皮建筑外观可保护建筑免受太阳直射。具有多个平面的外表皮与玻璃幕墙微微分开，将位于两者之间的检修通道隐藏起来。

金色的建筑立面在阳光下宛如精雕细琢的璀璨黄金，散发着耀眼的光芒，使之成为无可复制的视觉中心。

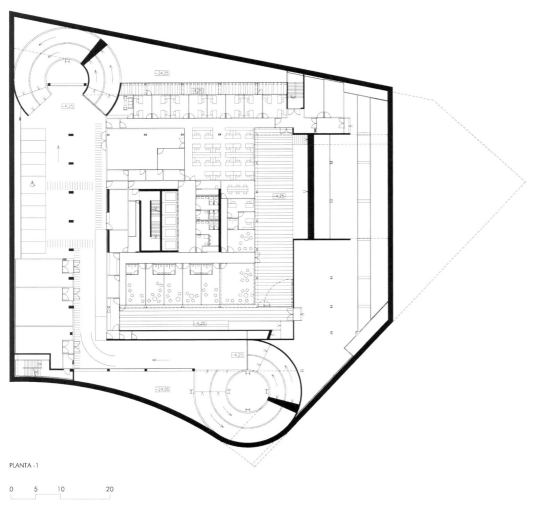

PLANTA -1

0 5 10 20

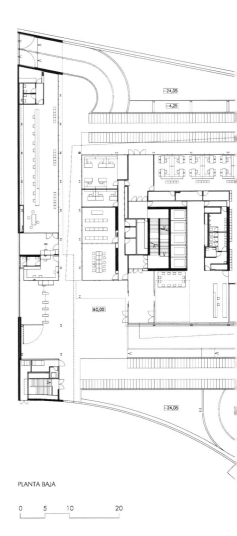

PLANTA BAJA

0 5 10 20

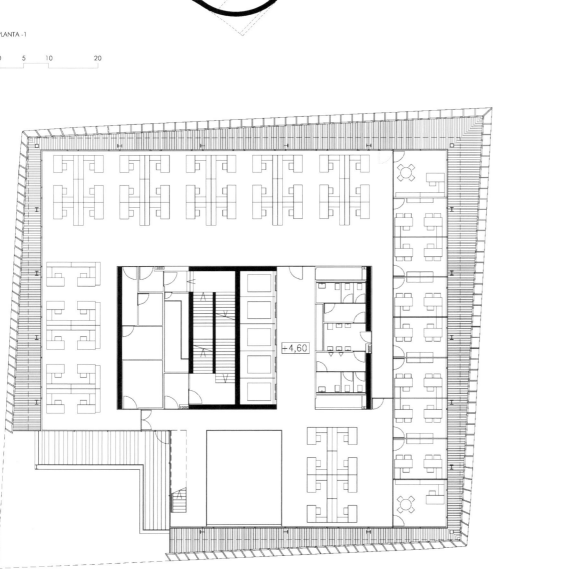

PLANTA 1

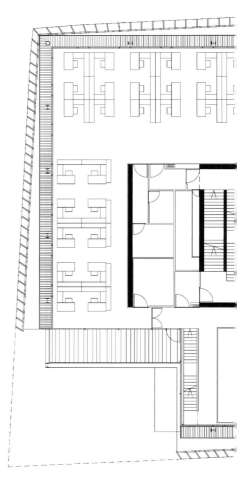

PLANTA 2

0 1 5 10

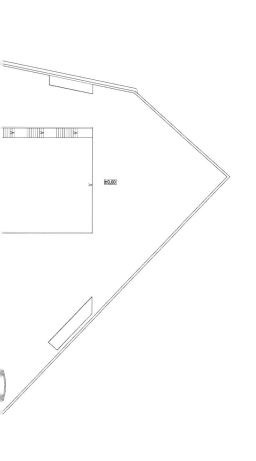
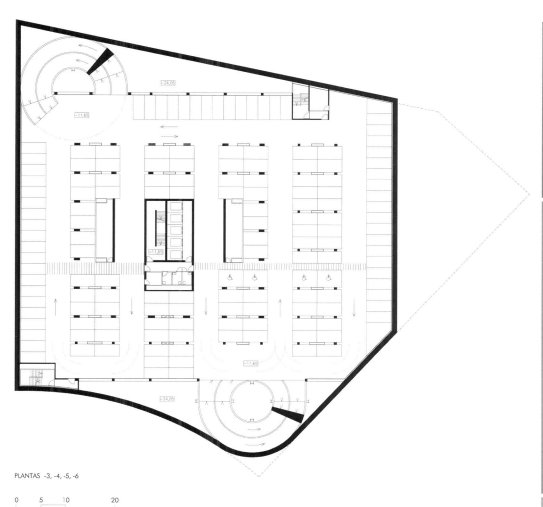

PLANTAS -3, -4, -5, -6

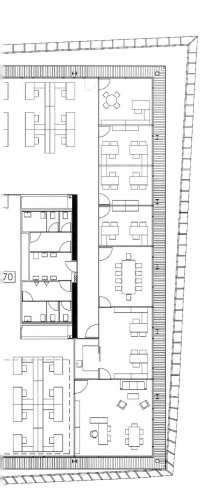
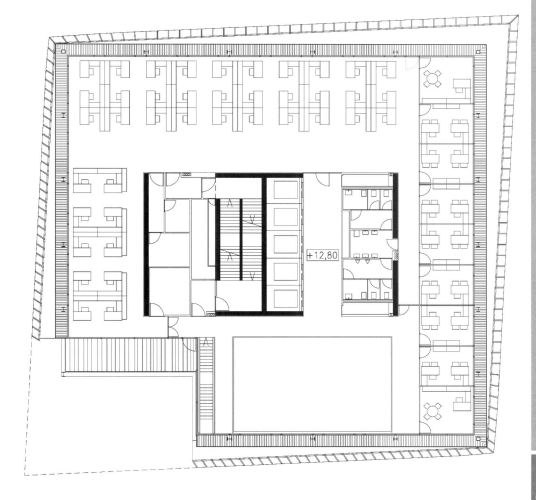

PLANTA 3

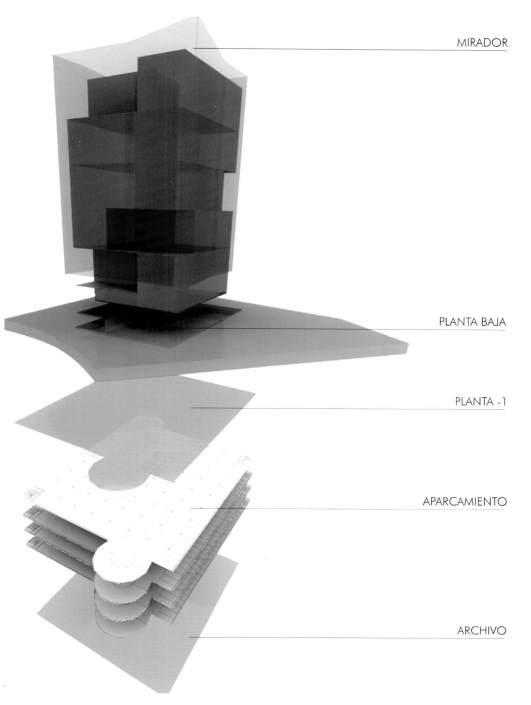

MIRADOR
PLANTA BAJA
PLANTA -1
APARCAMIENTO
ARCHIVO

Design Concept

The plot is located on the outskirts of Cordoba, on the E-W rail axis, the new backbone of the contemporary city. The existing buildings are devoid of any identity-related reference. Therefore, the construction of an institutional building for the Junta de Andalucía may well be the opportunity to give this sector a unique landmark building, with a self-contained, precise form.

The administrative building includes five Provincial Offices, various multi-purpose common areas and parking places. To embrace multiple functions for this administrative building, the proposal endows each unit with a unique and specific character, yet integrates them into a whole, where the whole must be greater than the sum of its parts.

Design Feature

The Provincial Offices will be housed in three successive levels. The lower level of each Office will consist of a large, triple-height foyer, a space for communication, sitting etc., open to a garden located at the corner of the floor. A light metal stairway will be placed between the foyer and garden to facilitate internal communication and ensure visual connection between Offices and the garden.

The modular office space allows a wide variety of configuration options. The flexibility and versatility of the interior space requires a neutral, uniform envelope to avoid jeopardising these opportunities. To that end, façades will be covered by a glass envelope, which in turn is protected by a metal lattice envelope to prevent direct sunlight. This envelope of faceted planes, slightly separated from the glass façade, conceals a light maintenance walkway located in the space between.

Golden façade in the sun resembles exquisitely carved gold emitting glaring lights and becoming incomparable visual center.

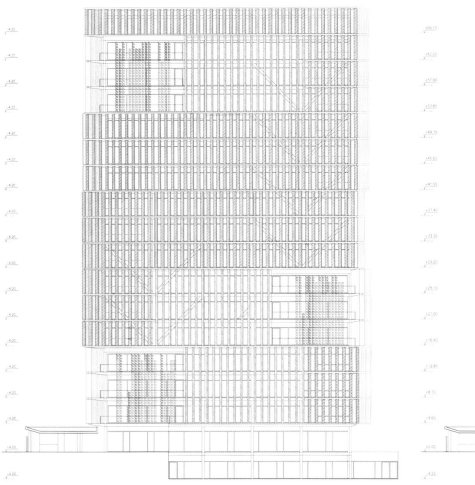

FACHADA SURESTE

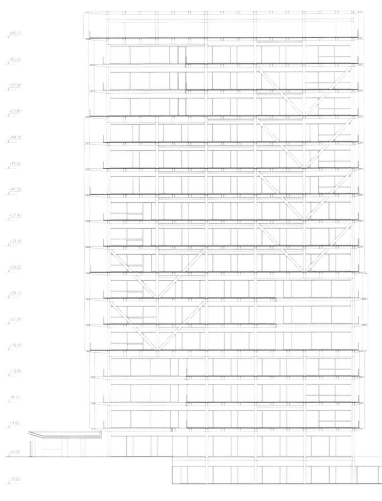

FACHADA SURESTE

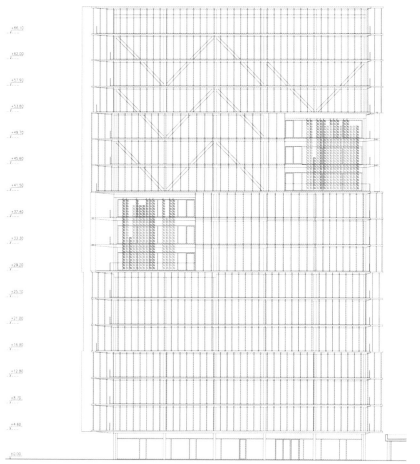

FACHADA NOROESTE

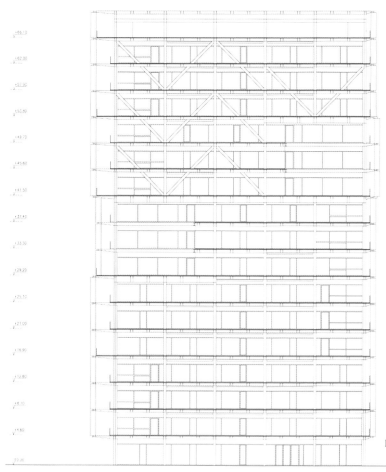

FACHADA NOROESTE

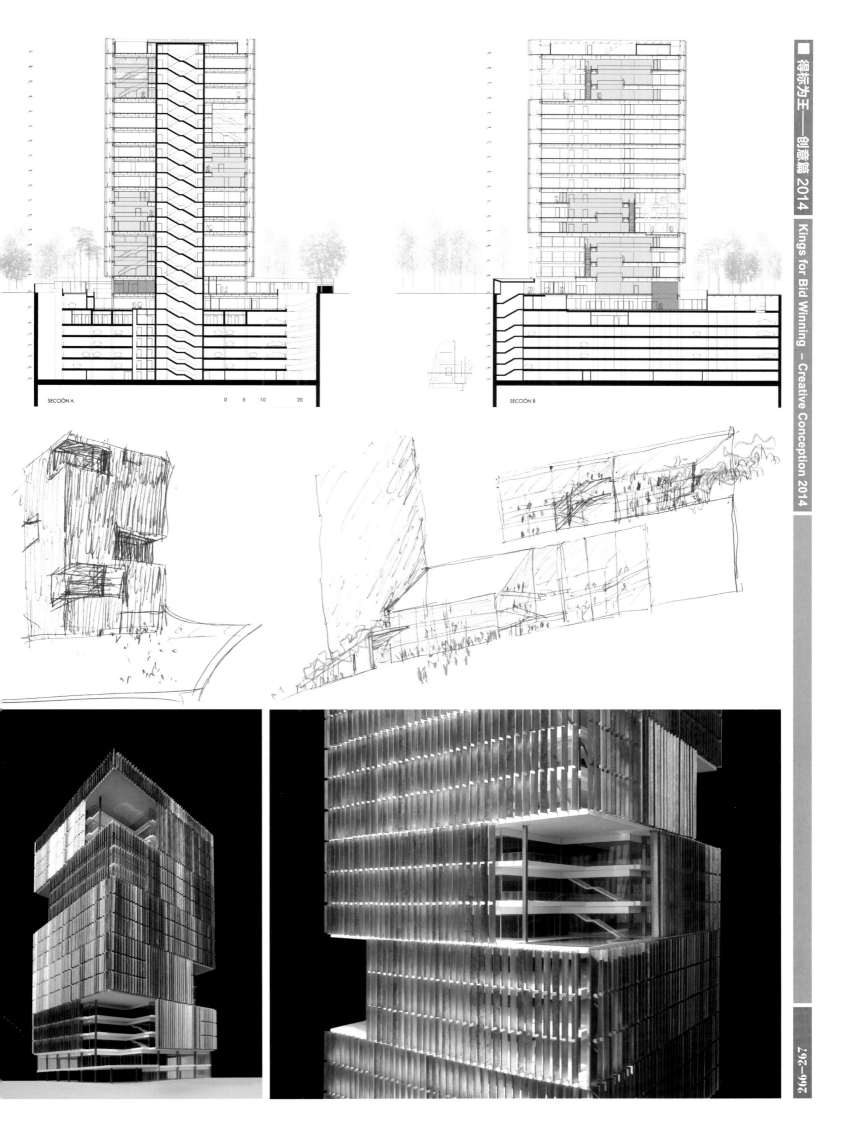

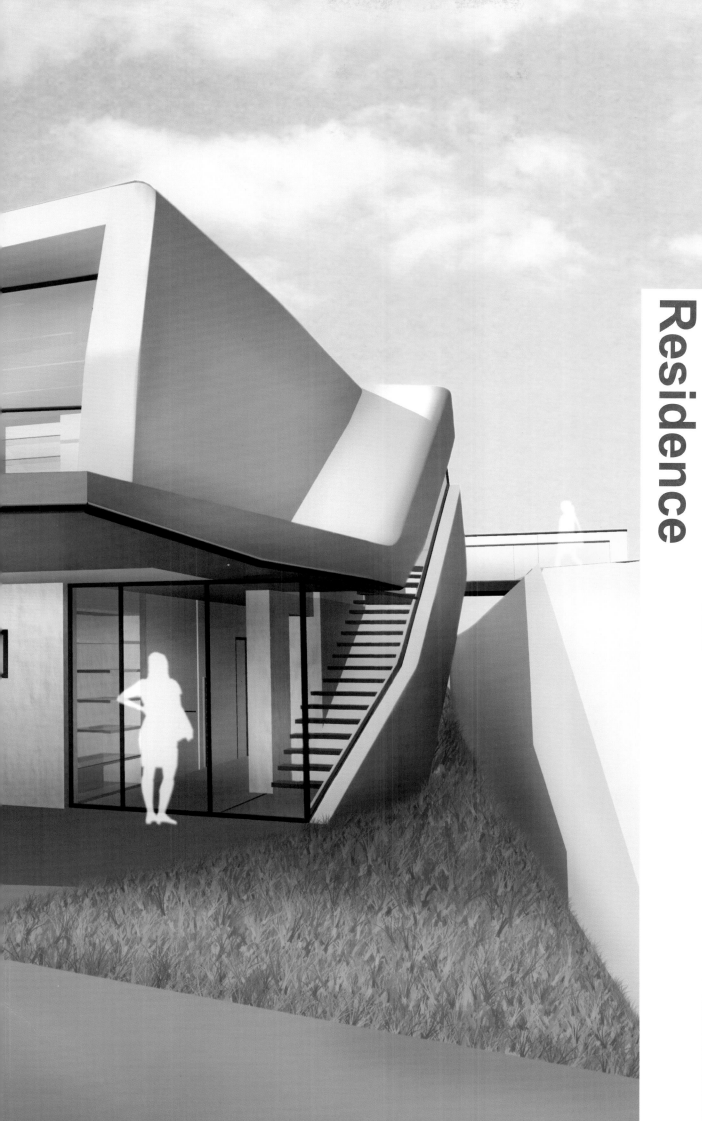

住宅建筑 Residence

几何体量
再生铝

Geometric Volume, Re-cycled Aluminum

瑞典哥特堡 Harbor Stones 住宅区
Harbor Stones

设计单位：C. F. Møller Architects
开发商：Norra Älvstranden Utveckling AB
　　　　PEAB
　　　　Stadsbyggnadskontoret
项目地址：瑞典哥特堡
项目面积：70 000 ㎡

Designed by: C.F. Møller
Client: Norra Älvstranden Utveckling AB;
PEAB; Stadsbyggnadskontoret
Location: Gothenburg, Sweden
Area: 70,000 m²

设计构思

项目的主要目标是为位于哥特堡旧工业港口 Lindholm 地区的新住宅区进行总体规划。项目的总体规划旨在在小规模城市空间引入符合人类尺度的多元化建筑类型，并明确建筑定位、建筑形态和透明度。

建筑设计

整个项目包括 10 栋几何形态的建筑体，其形态强化了"Harbor Stones"的象征意义。建筑高度在 4～10 层之间浮动，另外一栋 22 层高的塔楼则成为区域的地标。建筑立面采用了再生铝，呼应了 Lindholm 地区曾是一个造船厂聚集之地的历史。

在设计过程中，设计师重点考虑了吸收太阳光和扩大建筑视野两个方面。整个项目设有两座公共楼梯和一座景观楼梯，既构成了俯瞰海港的遮蔽空间，也为住户提供了开阔的景观视野。

Design Concept

The task has been to design the master plan for a new housing district in the Lindholm area of Gothenburg's old industrial port. The master plan proposal introduces a human scale and a different architecture focusing on the small-scale urban spaces within the scheme by refining the positioning, geometry and transparency of the buildings.

Architectural Design

The scheme consists of 10 sculptural building volumes. Their forms strengthen the symbolic meaning of "Harbor Stones", varying in height from four to ten stories. The scheme also includes a tower of approximately 22 stories which will rise as a landmark. The façades of buildings use re-cycled aluminum, making a strong statement with reference to the ship-building history of the site.

The buildings are designed to maximize the potential for sunlight and views of the harbor. Two public stairs and a third landscaped stair of the project create sheltered spaces directly overlooking the harbor and open views out for every resident.

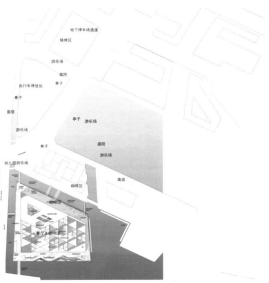

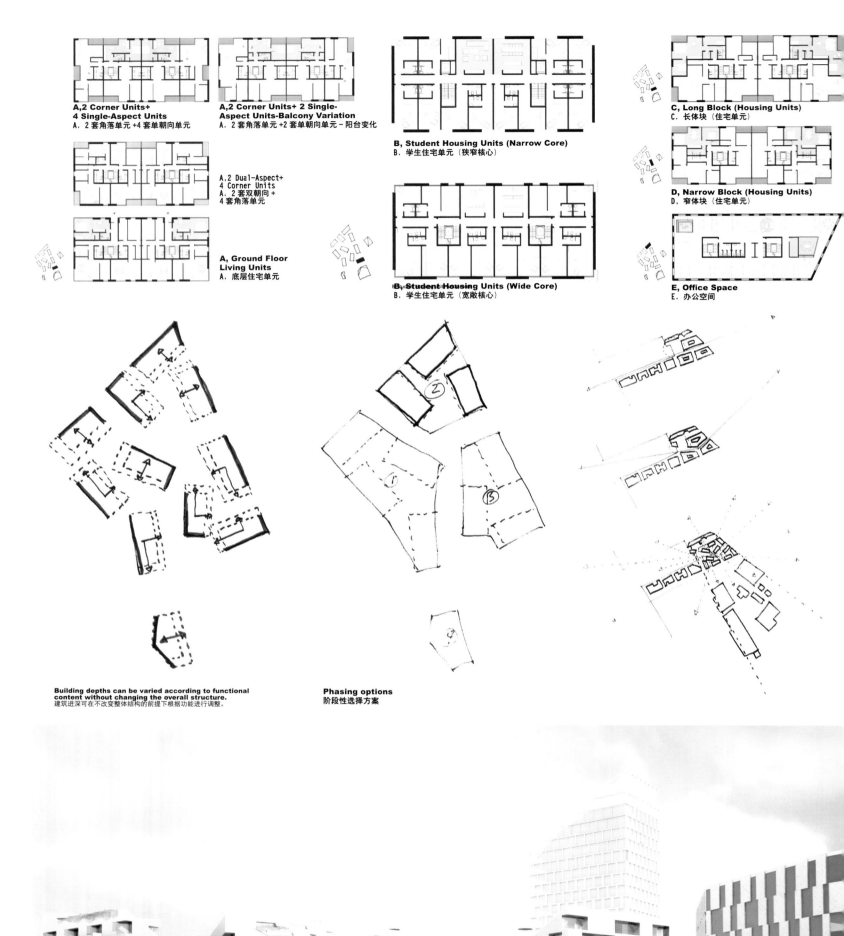

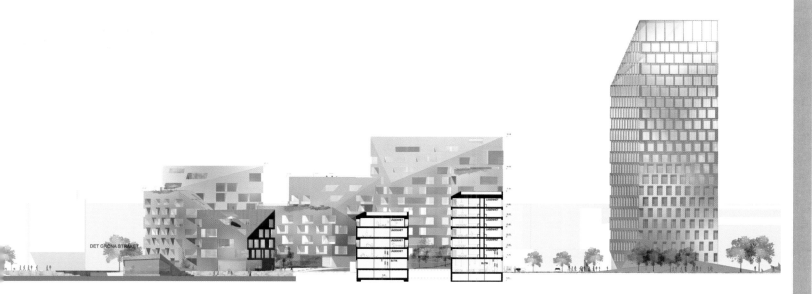

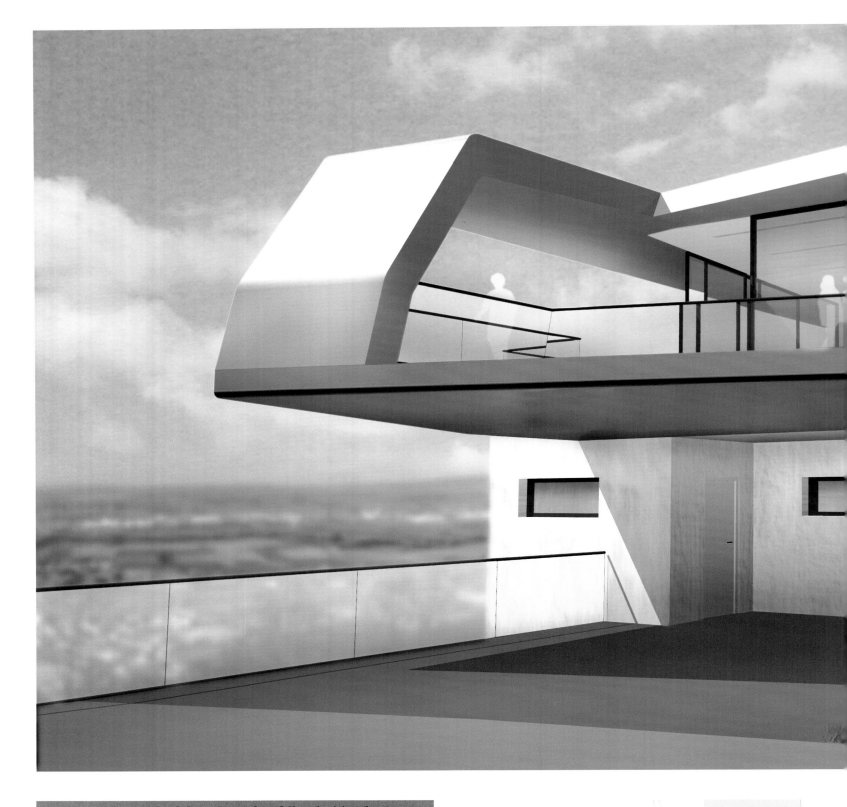

奥地利下奥地利州樱桃院住宅

Cherry Yard's House

设计单位：Architekt Lukas Göbl | Office for Explicit Architecture
项目地址：奥地利下奥地利州
项目面积：300 ㎡
设计团队：Lukas Göbl Oliver Ulrich

Designed by: Architekt Lukas Göbl | Office for Explicit Architecture
Location: Lower Austria, Austria
Area: 300 m²
Design Team: Lukas Göbl, Oliver Ulrich

"Turret", Sharp Contrast, Modern & Elegant

"角塔" 极致反差 现代典雅

设计构思

很多时候，极度反差的事物融合在一起，也能形成一种极致的和谐，就像光与影、黑与白、明与暗一样。本方案是对位于奥地利下奥地利州克莱姆斯地区一栋建于20世纪60年代的家庭住宅进行改造的设计方案，设计师将与原有建筑风格迥异的结构插入其中，使两者相互作用并融合，从而在功能和形态上达到一种独特的平衡。

设计特色

新旧部分在结构和风格上的对比因对建筑和外部材料的刻意分化而得以进一步加强。对于旧建筑，设计师对其隔热区进行改造，为其增添了一个古典风格的白色糙面外观，看起来十分古朴。新建部分则是以预制木材为主要元素的轻质结构，部分辅以钢架结构。屋顶、建筑立面、底面以及有屋顶遮盖的停车空间都覆盖了一个特殊的橡胶密封层，这一新型的建筑材料以典雅的灰色装扮了这一空间，赋予建筑现代时尚的气息。

两者的互补性基于主导性的功能理念而展开，这首要表现在新建部分补充并扩展了原有结构。新建结构容纳了所有的通道，并设有一个侧楼梯，建立了3个楼层之间的直接联系。分别向东和向南延伸的露台及类似的"角塔"结构界定了屋顶露台的范围，同时也营造了一个独特的室外空间。而从地板到屋顶的玻璃墙则为建筑提供了朝向克莱姆斯地区及南部克莱姆斯山谷的视野。

Design Concept

Many times, extremely contrasting elements fused together may create extreme harmony, such as light and shadow, black and white, dark and bright. The project is a renovation and expansion of a family home in Krems which was built in 1960s. The new form with very different structure and style blends with the old form, which achieves a unique balance of functions and forms.

Design Feature

The contrast between the new and old forms on structure and style is enhanced by the conscious differentiation of building and façade materials. The old building will be thermally renovated and given a classic, white, rough plaster exterior. The new building, on the other hand, is a lightweight construction erected mainly using prefabricated timber elements with partial steel supports. The roof, the façades, and the various undersides and projecting roofs covering the parking spaces are coated in a special rubber sealing layer which dresses the building in elegant gray.

The two structures are bound together through the overriding functional concept. The new building, which completes and "triggers" the old building, houses all pathways. A side stairway is also housed in the new building, creating a direct connection between all three floors. Two terraces extend east and south, and a sort of "turret" marks the end of the covered terrace and creates a very special outdoor place. Floor-to-ceiling glass walls frame the view of Krems and the southern Krems Valley.

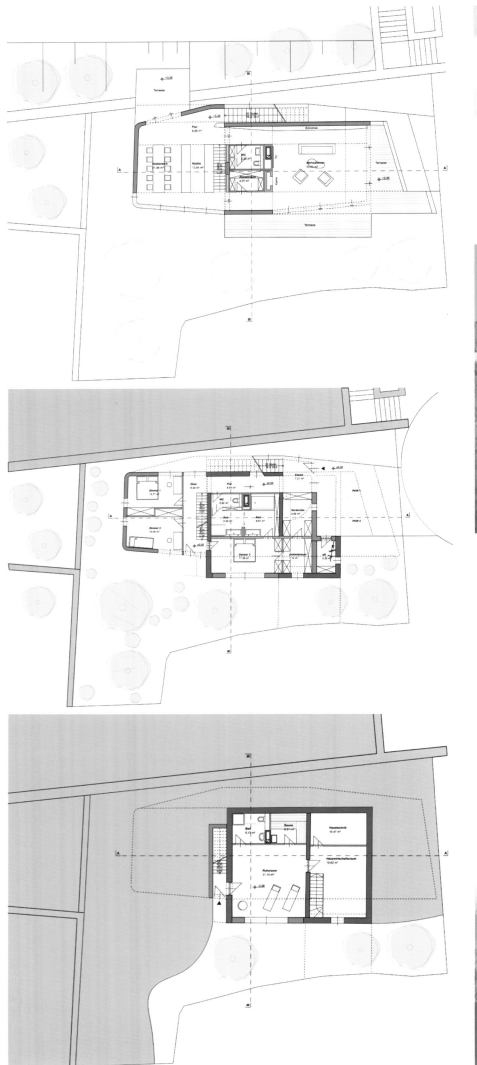
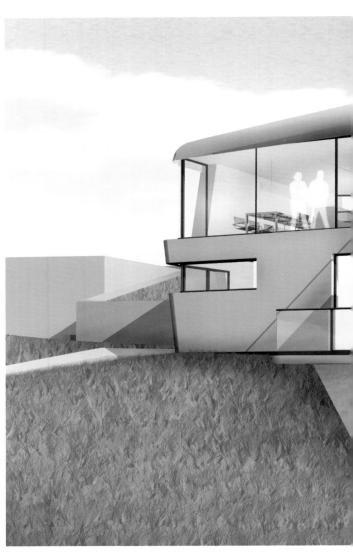
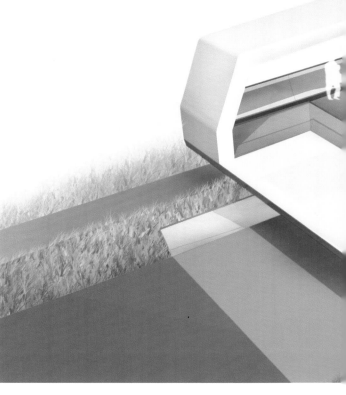

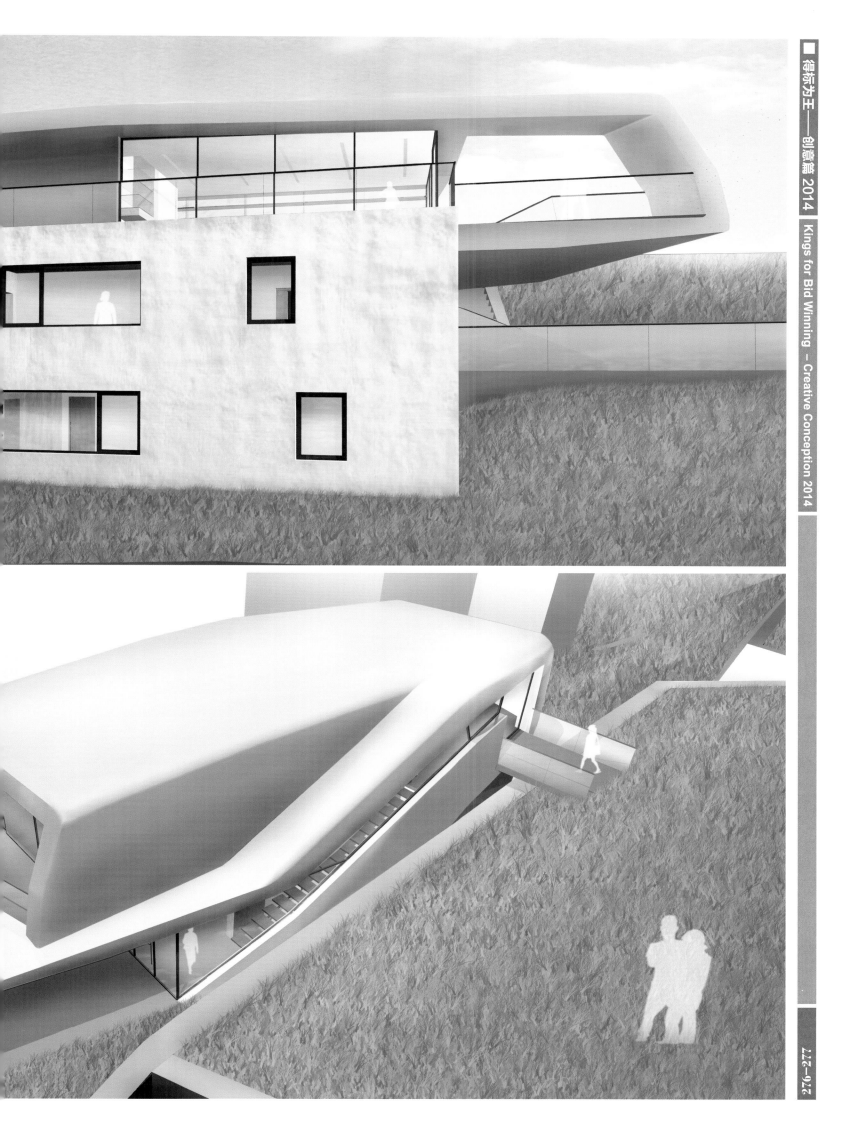

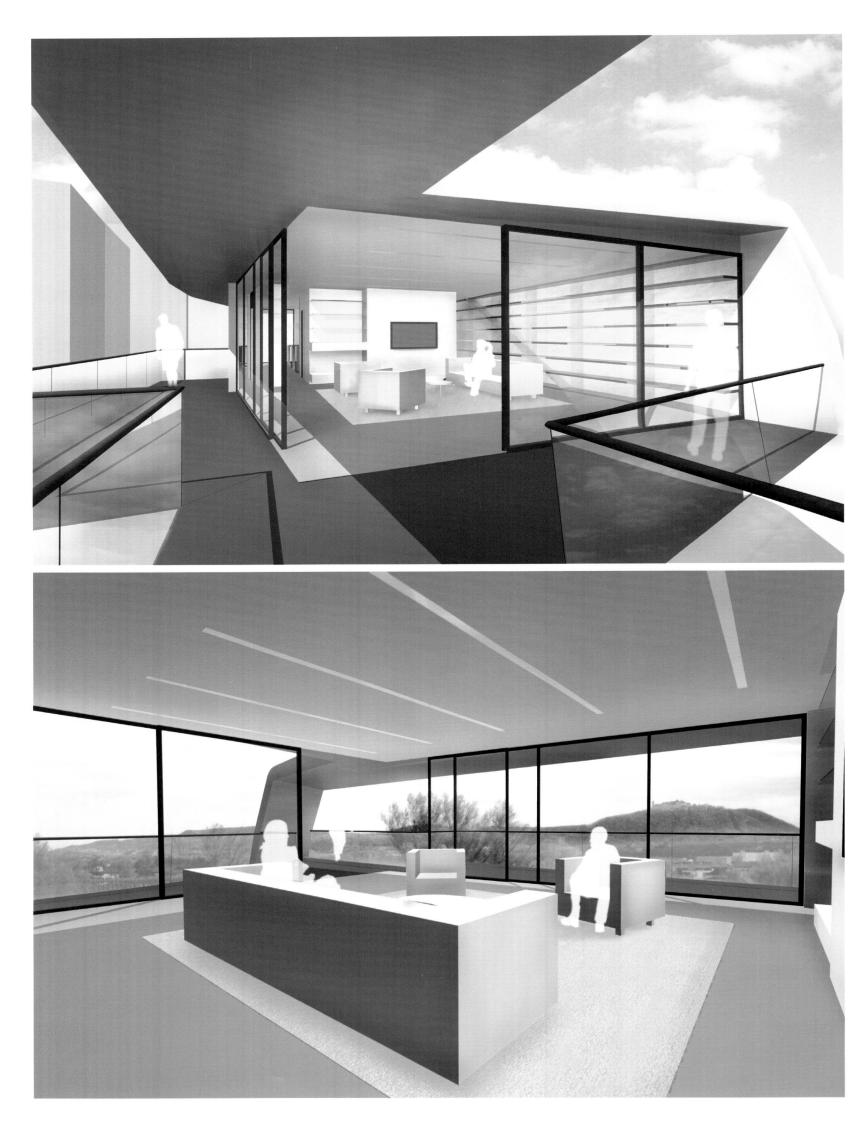

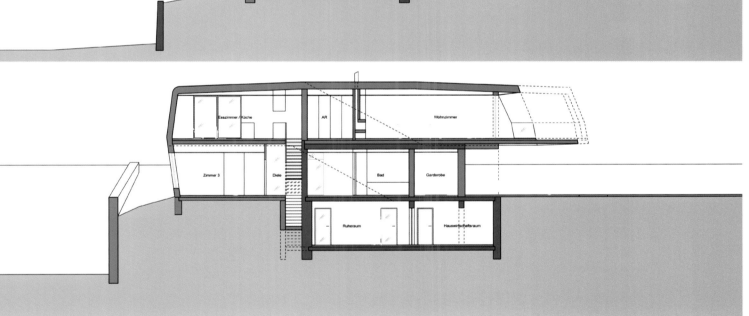

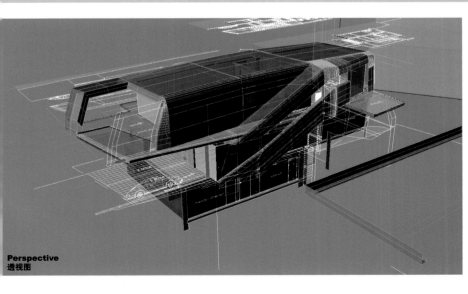

Perspective
透视图

瑞典斯德哥尔摩 Alvik 大楼
Alvik Tower

设计单位：C. F. Møller Architects	Designed by: C.F. Møller
开发商：PEAB Sverige AB	Client: PEAB Sverige AB
项目地址：瑞典斯德哥尔摩	Location: Stockholm, Sweden
建筑面积：20 000 m²	Floor Area: 20,000 m²
设计团队：Mads Mandrup Hansen　Bränd Jan-Erik Mattsson　Mårten Leringe　Sara Nilsson　Rasmus Brønnum	Design Team: Mads Mandrup Hansen, Bränd Jan-Erik Mattsson, Mårten Leringe, Sara Nilsson, Rasmus Brønnum

设计构思

项目场址位于朝向斯德哥尔摩群岛的一个显著的地块上，是城市空间与自然景观区之间的一个缓冲区，这一独特的地理特征激发了在这一地块内建造一个住宅综合体的设计灵感，特别是梅拉伦湖旁边险峻的山体形态为这个项目的塑形和线条勾勒提供了参照。

建筑设计

整个项目除了一个9层的断裂式体量和21层的标志性塔楼，还包括商店、餐厅和绿色空间。同时，方案为未来的开发设计预留了自由规划区间，这些空间可作为滨水区的观景平台，也可成为面向城市的中央广场，从而将项目与城市和街道平面联系起来，在为 Alvik 地区提供新的功能和身份特征时，将现代生活融入连续、活跃、富有吸引力的城市环境中。

在可持续性设计方面，建筑的形态可确保每一个住宅单元拥有最佳的采光条件，同时，建筑绝佳的热性能以及对太阳能的利用，也将提高建筑的能源利用率。

山体形态 断裂式体量

Steep Mountain, Fractured Block

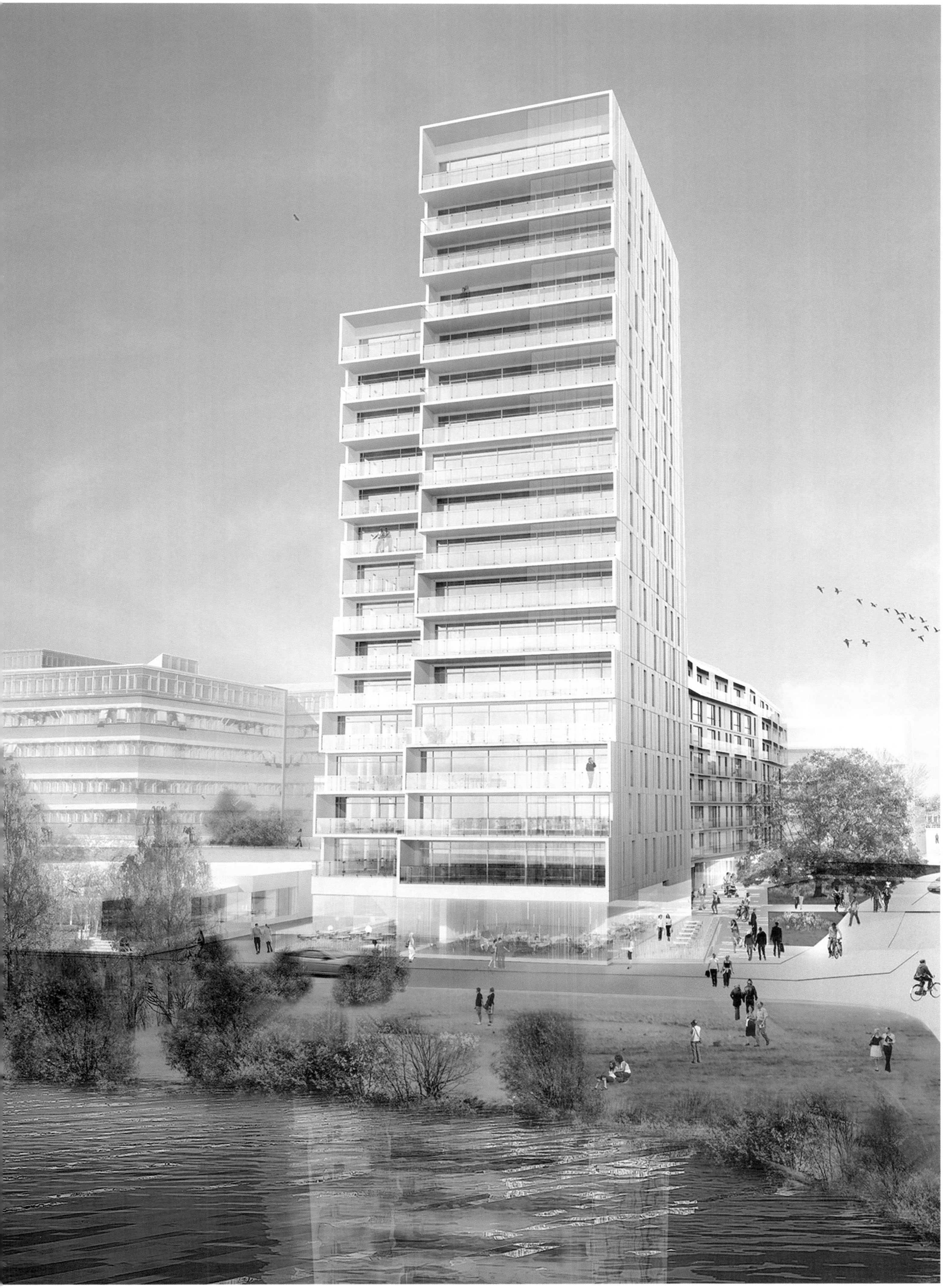

SITUATIONSPLAN 1:2,000
平面布置图 1:2 000

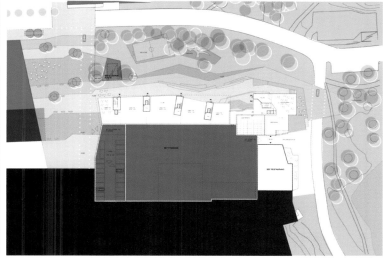

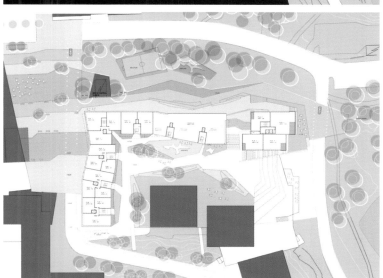

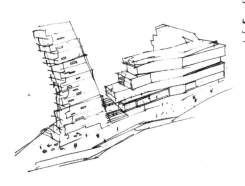

Design Concept

The site lies in a prominent location towards the archipelago, and forms a transition zone between the city and the landscape, which has inspired the architecture. The development draws its shape and lines from the steepness and structure of the mountains bordering the lake Mälaren.

Architectural Design

Except a fractured block of approximately nine stories and a landmark 21-story tower, the project also contains shops, a restaurant and green spaces. It provides free areas which can be used as squares or open spaces – such as a viewing level near the water, and a central square oriented towards the city. The development provides the Alvik area with new functions and identity, and creates a modern living compound in a continuous, attractive and active urban environment.

As to the sustainable design, the buildings' forms secures each residence a maximum of daylight and view. Excellent thermal performance and on-site solar energy generation also helps to improve the building's energy efficiency.

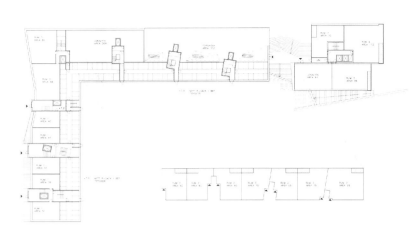

PLAN 01 1:500
平面图 01 1:500

PLAN 04
平面图 04

TYPE A - 2 værelser
1:200

TYPE A - 3 værelser
1:200

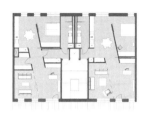

TYPE A1 - med gesims mod syd
3 værelser

TYPE A2 - med gesims mod nord + altan mod syd
3 værelser

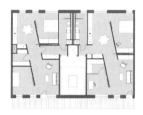
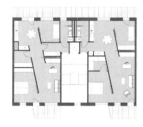

TYPE A1 - med gesims mod syd
3 værelser

TYPE A2 - med gesims mod nord + altan mod syd
3 værelser

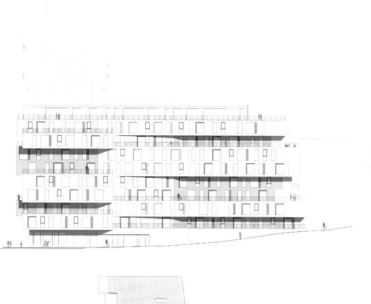
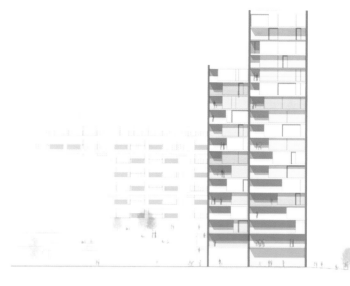
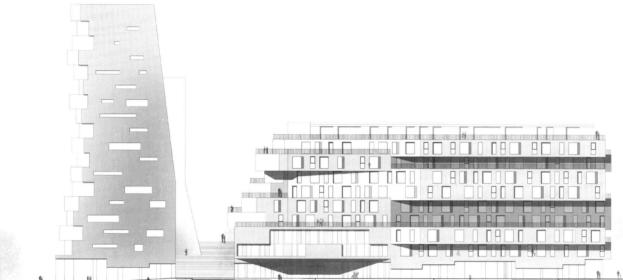

山脉
绿色公园
非常规设计

Mountain Range,
Green Park,
Unusual Design

荷兰皮尔默伦德 Londenhaven 公寓大楼

Londenhaven

设计单位：de Architekten Cie.
开发商：Bouwcompagnie, Hoorn
项目地址：荷兰皮尔默伦德
设计团队：Frits van Dongen　E. Thijssen
　　　　　N. Thomas　　　　N. Kozyra
　　　　　M. Paneque

Designed by: de Architekten Cie.
Client: Bouwcompagnie, Hoorn
Location: Purmerend, Netherlands
Design Team: Frits van Dongen, E. Thijssen,
N. Thomas, N. Kozyra, M. Paneque

项目概况

Londenhaven 公寓大楼将是皮尔默伦德市 Europa 地区住宅开发项目中的佼佼者，设计提出了一种全新的设计理念，即以城市规划的方式来实现建筑各方面的职能。

建筑设计

这个多样化的住宅项目包括独户住宅、一栋狭长的公寓大楼以及一个护理中心，其中，公寓大楼因其在高度上的独有特征成为一个地标式建筑，也是通往皮尔默伦德城区的入口标志。Europa 地区的公寓大楼都设置有庭院空间，Londenhaven 公寓大楼的设计遵循了这一格局，其庭院空间占地近 1 公顷，宛若一个小型的绿色公园。

为解决项目的停车问题，设计采取了非常规的途径：公寓大楼的住户将车停在建筑下方的停车场，该停车场直接与街道平面连通，使公寓大楼可不受交通的影响；另一方面，独栋住宅可将车停泊在各自的物业区，从公寓大楼处的道路出入。

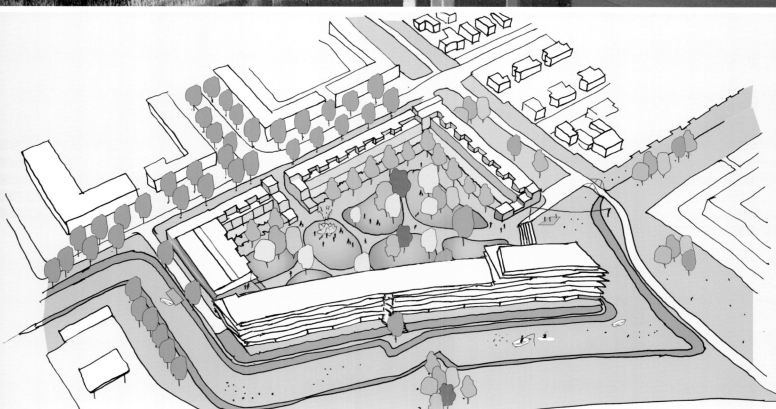

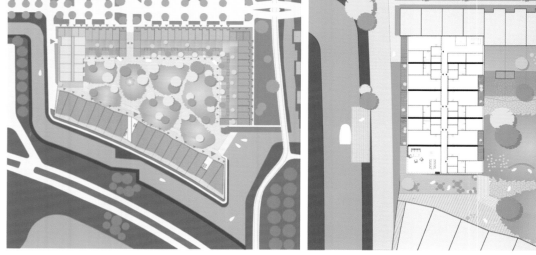

Profile
The city court Londenhaven is the primus inter pares among the courts in the new Europa Quarter extension to Purmerend. A powerful concept has been developed, realizing the potential of the city court as an urban planning typology in all respects.

Architectural Design
The diverse housing program comprises single-family houses, a care center and an elongated apartment building. The apartment building with its distinctive height acts as a landmark, marking the entry to the town of Purmerend. Each court of Europa Quarter has a yard with a character of its own. The courtyard of Londenhaven covers an area of almost one hectare. It is green and will breathe the atmosphere of a public park.

For parking an unusual solution is developed: most cars are parked in a garage under the apartment block, directly accessible from the street, thus keeping the court free of this traffic. The inhabitants of the single-family houses park their cars on their own property, accessed through an informal route through the court.

BG   **1**   **2**   **3**

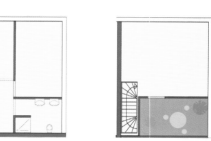

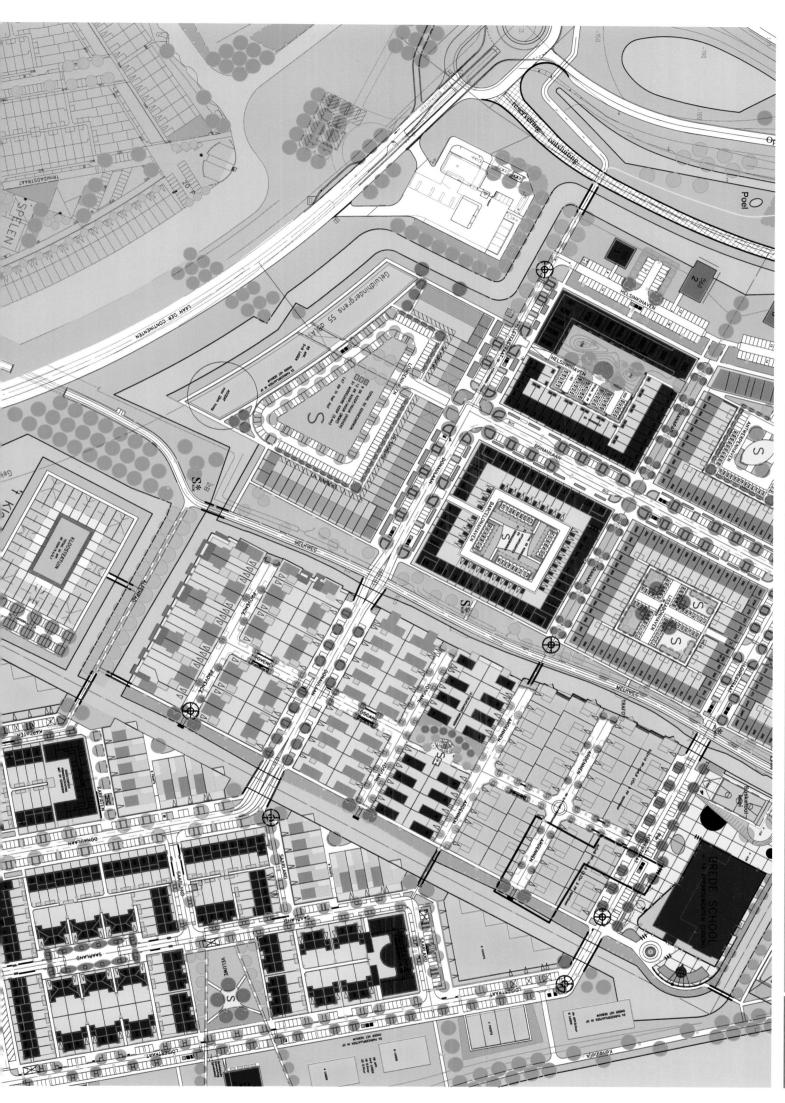

秘鲁利马精品公寓
Loft Boutique

设计单位：TheeAe LTD.
项目地址：秘鲁利马
基地面积：2 700 ㎡
建筑面积：7 505 ㎡

Designed by: TheeAe LTD.
Location: Lima, Peru
Site Area: 2,700 m²
Floor Area: 7,505 m²

Natural Form, Private Space

树叶肌理
私密空间

项目概况

这一精品公寓是为向往生活在圣伊西德罗最酷地区的单身群体量身打造的。考虑到该区域可建造区域对场地的限制性因素，这一住宅大楼定义为 16 层，共包括 30 个住宅单元。

设计特色

考虑到项目的基本定位和场地的限制性因素，建筑的形态和布局尽可能地利用周边可建造区域的环境，使建筑尽可能地获得朝向公园和利马高尔夫俱乐部的景观视野。这样一来，建筑形态形成了朝向公园的定向流动纹理，外立面上源自大自然的、具有一定方向性的树叶状肌理，既加深了这一视觉效果，也使建筑在大都市间流露出一丝天然的气息，呼应了现代人内心中对自然的向往。

整个居住单元包括 30 套单体公寓和复式公寓，建筑的形态和布局则确保了每一个住宅单元都朝向利马高尔夫俱乐部。这些单元的室内净高最小为 2.7 m，并为厨房、洗衣房和每一个房间提供一整套完善的煤气设备，方便单身群体的日常生活。另外，厨房设置直接的通风设施和照明设备，洗衣房则通过室内采光井自然通风。建筑最顶层的屋顶露台花园为住户提供了一个清幽、静谧、不被打扰的休闲场所，从心理诉求方面满足单身群体对自由、私密生活的需求。

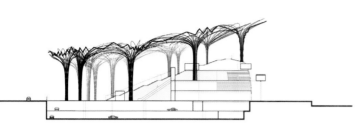
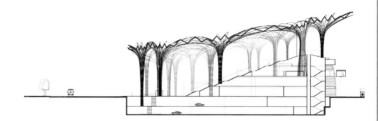
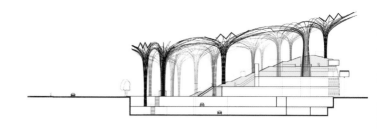

CURVED FLOOR PATTERN WITH NATURAL STONES PAVING
The natural pattern enhances the design of the new residential tower in Lima.
自然石铺装曲线地板模式
自然图案突出了利马新住宅大楼的设计。

CORE TOP ENVIRONMENTAL FRIENDLY
Provided natural plants on the roof will reflect building design for environmental friendly.
核心顶级环保
屋顶的自然植物将反映出建筑设计环保的一面。

OUTSIDE GRILL
Provided grills will be functioned well for the resident to get together to invite friends and families.
户外烧烤架
提供烧烤架以便于居民与亲人及朋友聚会。

EMERGENCY EXTERNAL STAIR CASE
Space for external staircase has been allocated for applicable local code provision to meet two means of egress.
应急外部楼梯
根据当地的规定,设计了外部楼梯空间以满足两种出入方式的要求。

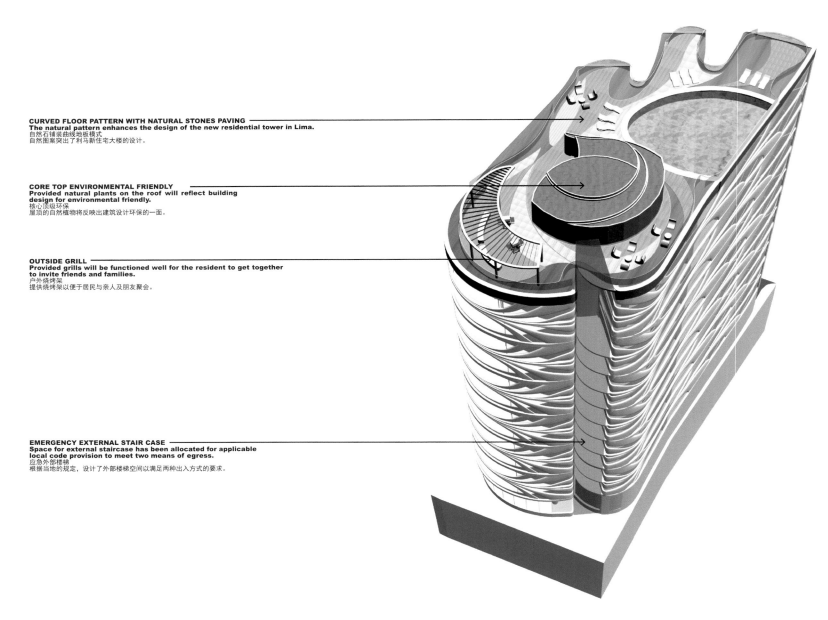

TYPICAL FLOOR PLANS AT LEAVE 2,4,6,8,10&12
2.4.6.8.10&12 层标准层平面图

Unit A
单元 A
3 Bedrooms
3 个卧室
Living Room
客厅
Dining Room
餐厅
2 Full Bathroom+1/2 Bathroom
2 个带浴室的卫生间 + 浴室

Unit B (duplex)
单元 B (复式)
3 Bedrooms
3 个卧室
Living Room
客厅
Dining Room
餐厅
2 Full Bathroom+1/2 Bathroom
2 个带浴室的卫生间 + 浴室

Unit C
单元 C
3 Bedrooms
3 个卧室
Living Room
客厅
Dining Room
餐厅
2 Full Bathroom+1/2 Bathroom
2 个带浴室的卫生间 + 浴室

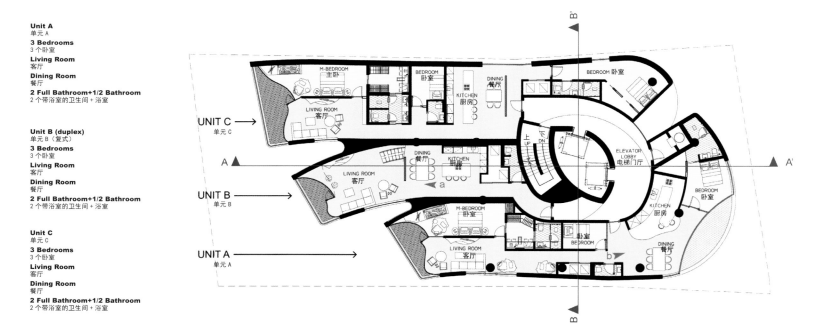

PENTHOUSE UNIT VS TYPICAL FLOOR UNITS
阁楼单元 VS 标准层单元

Units are consist of 3 different types for 27 units +3 Penthouse units additionally. Plans are developed based on typical layout as well as penthouse layout. The difference between typical unit and penthouse is the ceiling height. Because of swimming pool requirement depth, the ceiling height automatically generated. By doing that, it has been economically developed maintaining major façade element to be repetitive, yet it shows great variations of design.

27 套 3 种不同类型的单元 +3 套阁楼单元。建筑平面设计以标准层布局以及阁楼布局为基础。标准单元和阁楼单元的不同点在于室内净高。依据游泳池所要求的深度,自动生成了室内净高。这样,建筑实现了经济节约型开发,主要立面元素重复出现,同时实现了设计的多变。

FLOOR PLANS ISOMETRIC NW- SCALE 1/200
楼层平面等距离西北 - 比例 1/200

1.1. Typical Level 3,5,7,9,11,& 13 Floor Plan.
-Unit A&C
-Unit B (Duplex- Upper Floor)

1.1. 3, 5, 7, 9, 11 & 13 标准层
楼层平面图
-A 单元 &C 单元
-B 单元（复式上层楼面）

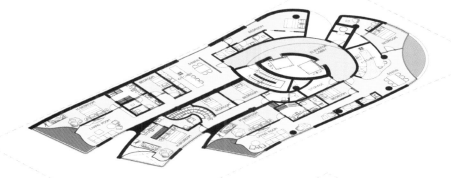

1.2. Typical Level 2,4,6,8,10 & 12 Floor Plan.
-Unit A&C
-Unit B (Duplex-Lower Floor)

1.2. 2, 4, 6, 8, 10 & 12 标准层
楼层平面图
-A 单元 &C 单元
-B 单元（复式下层楼面）

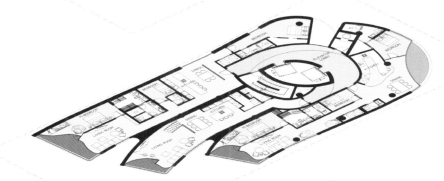

1.3. Level 1 Floor Plan
- Administration Office
- Mail Room
- Residential Entrance Lobby
- Starbucks Coffee Shop

1.3. 一层平面图
- 行政办公室
- 收发室
- 住宅入口大厅
- 星巴克咖啡店

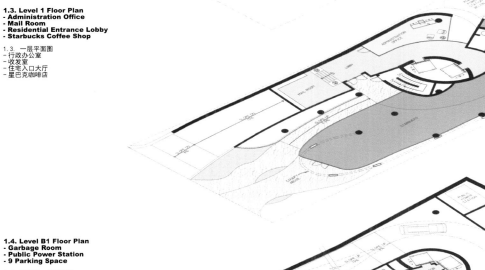

1.4. Level B1 Floor Plan
- Garbage Room
- Public Power Station
- 9 Parking Space

1.4. 地下室一层平面图
- 垃圾房
- 公共发电站
- 9 个停车位

1.5. Level B2~B4 Floor Plan
- 15 Parking Space at Each Floor

1.5. 地下室 2 层 -4 层平面图
- 每个楼层 15 个停车位

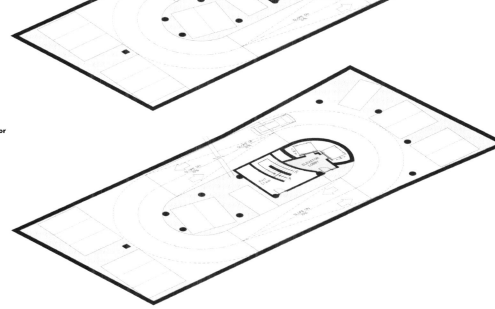

Loft Boutique 2010
阁楼精品 2010

Tower in Front of Golf Club, Lima Peru
秘鲁利马高尔夫俱乐部的塔楼

Building Description
建筑说明

Building Usage: Residential
建筑用途：居住

Floors/ Height: Total 17F (Basement 4)/ 48m
楼层 / 高度：共 17 层（地下室 4 层）/48m

Residential Floor Area: 5,922sm
居住楼层面积：5 922 ㎡

Lobby and Service Area: 1,312sm
大厅和服务区：1 312 ㎡

Office + Mail Room: 61sm
办公室 + 收发室：61 ㎡

Starbucks: 210sm
星巴克：210 ㎡

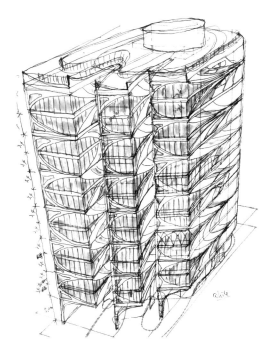
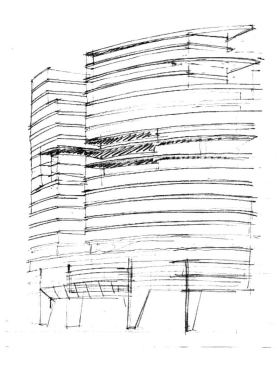

Pattern Elements
图案元素

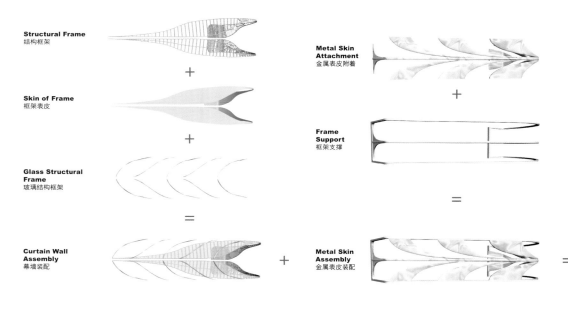

Structural Frame
结构框架

Skin of Frame
框架表皮

Glass Structural Frame
玻璃结构框架

Curtain Wall Assembly
幕墙装配

Metal Skin Attachment
金属表皮附着

Frame Support
框架支撑

Metal Skin Assembly
金属表皮装配

Assembly of Curtain Wall
幕墙装配

Building curtain wall is divided into five different elements. Each elements are connected each other in order to support additional structural stability.
建筑幕墙分为 5 个不同的元素。5 个元素互相连接以增强结构的稳定性

Skin of frame will be utilized for various shapes to control support area based on structural requirement by engineers.
根据工程师的要求，框架表皮将用于不同形态中以控制辅助区。

PATTERN STUDY
图案研究

Directional Façade Pattern Originated from Core Location
定向型外观图案源自核心位置

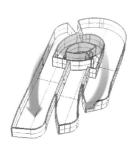

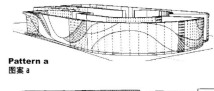

Pattern a
图案 a

Pattern b
图案 b

Pattern c
图案 c

Pattern d
图案 d

Final Directional Pattern Idea
最终定向图案理念

Core arrangement by building program naturally allowed directional pattern toward Lima Golf View.
项目方案核心部分的布局使建筑拥有利马高尔夫俱乐部的定向型图案。

Directional Leave, such elements found in nature was involved to develop design ideas for pattern of building façade.
项目理念涉及了自然中可以找到的定向元素来营造建筑立面图案。

Sun Light Model Study
光照模型研究

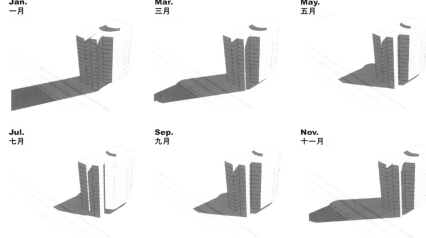

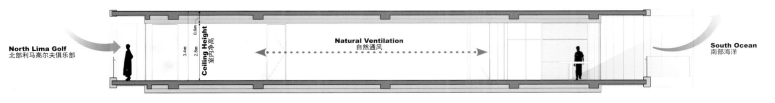

Typical Section at Residential Unit A
A 住宅单元典型剖面

Utilization of natural ventilation was in main concern to lay the balconies at both North and South of the unit. It blocks the Sun light from the south and provide open wide view toward Lima Golf with providing sound buffer zone from main traffic road from the north.
在住宅单元的北侧和南侧设置阳台以确保自然通风。同时，阳台阻隔了来自南侧的自然光，提供了面向利马高尔夫俱乐部的开阔视野，也提供了北侧主要交通道路与住宅之间的一个噪音缓冲区。

Plus, provided unit floor height 3.4m would be advantageous to realize the possible natural ventilation and lighting.
此外，3.4m 的楼层高度将有利于实现自然通风和采光。

Profile

The major use for the units is for a single person who wishes to live in the coolest zone of San Isidro. In addition, the site is limited by buildable area with height restriction which allows maximum 16 story height for 30 residential units.

Design Feature

For the consideration of the site's basic positioning and limitations, the design was initiated with a form to use the basic buildable area and updated after that for the most efficient viewable shape toward the park and Lima Golf Club. By doing so, the shape of building generated its directional flow toward the park, so the façade was created to increase its natural form as organic leave. It heightens visual impact, provokes natural atmosphere in the city whilst meets people's needs for nature.

Residential units consisted of flats and duplex apartments with 30 apartments. All residential units are facing the Lima Golf Club and the clear ceiling height is at least 2.7 meters for each unit. Kitchens, laundries and each of the rooms are provided with an integral gas installation to facilitate the users' daily life. At the last level, the building includes a roof terrace-garden to offer a tranquil undisturbed leisure place for single ones to enjoy free private life.

Unit Arrangement in Elevations
立面上的单元布局

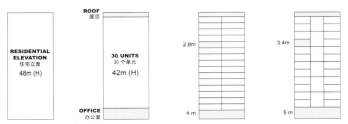

Unit Arrangement in Plans
平面上的单元布局

Parking Circulation
停车流线

墨西哥合众国瓦拉塔港 La Peninsula

La Peninsula

设计单位：Pascal Arquitectos
项目地址：墨西哥合众国瓦拉塔港

Designed by: Pascal Arquitectos
Location: Marina Vallarta, Mexico

设计构思

项目的设计基于两个基本理念而展开：一是为了打破当地已饱和的低端住宅格局，实现对场地的最大化开发利用；二是确保项目的商业价值，使其物有所值。为此，设计师从建筑的开放性、建筑与景观的关系以及功能配套着手，实现这两个设计目标。

建筑设计

设计师构想了两列带有大型花园、面积达 19 056 ㎡ 的建筑群，这些建筑体量包括公寓大楼、集中布局的服务区和配套设施。塔楼的较低楼层是开放式的，前3个公寓层则有着透明的外观，构成了一个开放型的发展项目。

设计十分注重建筑与景观的关系。考虑到场地面临海港的地理因素，设计师将建筑沿海湾的轮廓呈弧状排列，在场地内勾画出两轮背对而立的"新月"，既获得了最佳的景观视野，又与附近海滩和巴亚尔塔港建立了直接的视觉联系。

除了提供大型的公共花园区以外，该项目还为居民提供了多种辅助性服务，包括公共洗手间、仓库、行政办公室、机房、警卫室、服务用房、垃圾房等其他配套设施。

Design Concept

The project design is based on two concepts: one that allows a maximum exploitation of possibilities without falling on a typical saturated low range residence, and a second one, very important too, to obtain an acceptable investment return. To achieve the two design goals, designers emphasize on an open design, the relationship between architecture and landscape, and supporting functions.

Architectural Design

Two 19,056 m² buildings development with construction-free garden areas are created. These architectural bodies include the apartment towers and concentrate services and installations. Towers lower levels remain completely open a the first three apartment levels are transparent, which contribu to an open-type development project.

The design lays emphasis on the relationship betwee architecture and landscape. Since the site facing to the harb designers set buildings along the bay with an arc patter creating two "crescents" back to back in the site. It takes t best out of the gorgeous views whilst creates a direct visu contact with the nearby beaches and Marina Vallarta.

Besides extensive gardened areas, the development provides variety of complementary services for the inhabitants includi restrooms, warehouses, administrative offices, engine room guardhouses, service rooms, garbage rooms and other supp services.

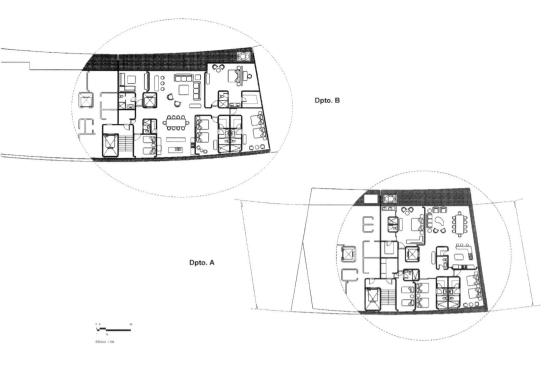
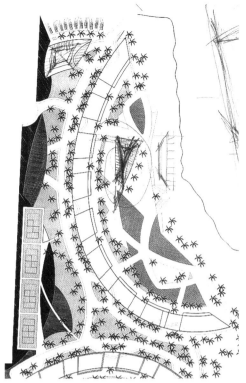

"新月"
弧形轮廓
景观最大化

"Crescents",
Arc Pattern,
Maximized Landscape View

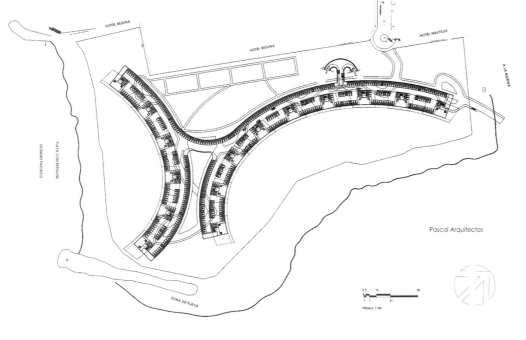
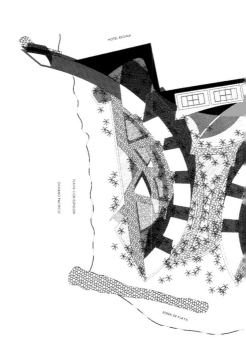

Pascal Arquitectos

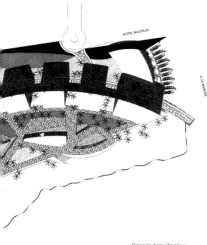
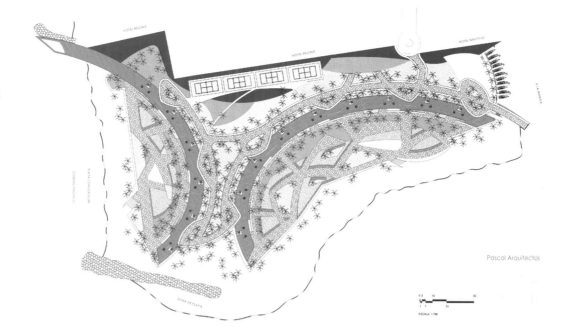

Pascal Arquitectos

Pascal Arquitectos

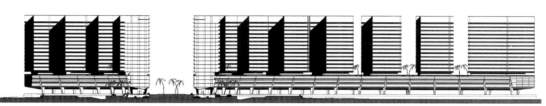
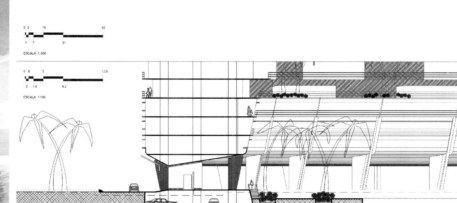

Pascal Arquitectos

综合
Others

台湾台中"台湾塔"
Taiwan Tower

设计单位：Casanova + Hernandez Architects
开发单位：台中市政府
项目地址：中国台湾省台中市

Designed by: Casanova + Hernandez Architects
Client: Taichung City Government
Location: Taichung, Taiwan, China

Plum Blossom, Geometric Structure, Tree-shaped Structure

梅花
几何体量
生命之树

项目概况

无论是从象征意义上还是实际意义上来说，这栋独立的建筑不仅将成为这座城市的视觉标志，同时也是台湾精神以及台湾可持续化进程的象征。

设计理念

自然中生长的植物为这个塔楼的结构设计提供了灵感，这一"树形"结构的建筑既可抵抗台风、地震等自然灾害的侵袭，同时也体现了可持续发展、人与自然和谐相处、尊重环境等理念，这使这一建筑成为一个典型的可持续性建筑。

设计特色

"台湾塔"的形态和功能组织皆源自象征台湾精神的梅花。建筑被分为5个部分，它们像5片花瓣一样组织在一起，各自朝向5个不同的方向，在5个不同的高度上为观赏城市景观提供了5个观景平台。

"台湾塔"好似一个垂直的绿色城市，包括城市街道、广场和与"台湾塔"底层公共广场相连的公园。这一建筑中设有餐厅、商店等娱乐设施，台中市开发区博物馆等文化建筑以及环境质量监测站和博物馆办公室等办公空间，这些功能区分布在不同的楼层。

这个几何形态的体量体现了均衡、和谐和可持续发展等自然主义特征，绿色植物平台、空中花园加强了这一特征，也使"台湾塔"的外表面成为对可持续性的回应。

"台湾塔"不仅仅是一个建筑，同时也是一个象征。这栋大楼从设计理念到细部处理中涵盖的大量元素，都具有具体的象征意义，这一象征主义的设计手法，将这栋大楼与台湾的过去、现在和将来联系起来。

Profile

Both aspects, the symbolic and the physical ones, have to merge in a single building to be not only a visual landmark in the city but, at the same time, a symbol of the Taiwanese spirit and its sustainable progress.

Design Concept

The structure of the tower is inspired by the natural growth of the plants in order to create a strong structure resistant to cyclones and earthquakes but at the same time capable of symbolizing values such as sustainable growth, natural harmony and environmental respect to become an icon of the sustainable architecture.

Design Feature

Taiwan Tower bases its spatial and functional organization on the Taiwan Flower, the Plum Blossom, by dividing the tower in 5 parts, organized as five petals, orientated in 5 different directions that offer 5 observation decks on five different heights looking over the city.

Taiwan Tower works as a Vertical Green City consisting of public streets, squares and parks that connect Taiwan Tower public plaza on the ground floor with recreational facilities such as restaurants and shops, cultural buildings like the Museum of Taichung City Development and working spaces such as the Environmental Quality Monitoring Station and the museum offices on different levels.

The geometry of the structure is based on natural values such as equilibrium, harmony and sustainability that, reinforced by vegetation in platforms and sky-gardens, makes from the Taiwan Tower exterior image an evoking icon of sustainability. Taiwan Tower is not only a building, but a symbol. That is why most of the elements of the tower from the architectural concept to the smaller details have a symbolic meaning. By this, the building will remain in the collective memory of Taiwanese people and foreign visitors associated to the past, present and future of Taiwan.

保加利亚索菲亚 Collider 活动中心
Collider Activity Center

设计单位：Process-based Architecture Studio	Designed by: Process-based Architecture Studio
开发商：Walltopia 公司	Client: Walltopia Company
项目地址：保加利亚索菲亚	Location: Sofia, Bulgaria
设计团队：Jafar Bazzaz, Arash Pouresmaeil, Kamal Youssefpoor	Design Team: Jafar Bazzaz, Arash Pouresmaeil, Kamal Youssefpoor

Outdoor Climbing, Elevated Ground Floor, Land of Happiness

户外攀岩
底层架空
休闲乐土

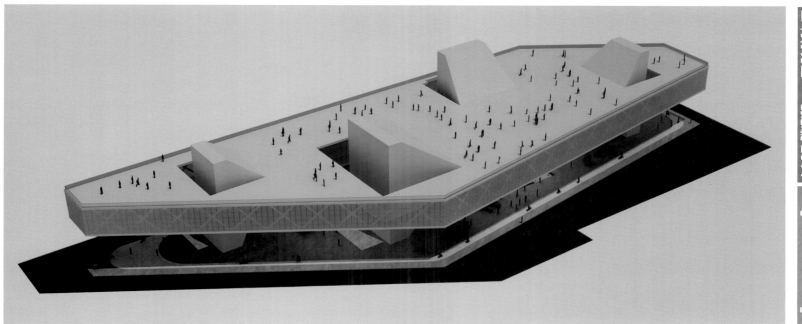

设计理念

毋庸置疑,这是一个极富创意的设想,这无关建筑的形态,而是这一方案为人们提供了一种新型的攀岩模式——一种介于自然攀岩与室内攀岩之间的户外攀岩,这一设想在很大程度上提高了建筑的开放性,使建筑与城市空间建立了更为直接的联系。

设计特色

Collider活动中心的主体结构是一个简约的大型立方体。建筑的体量被大幅度抬升,空间的四周都是开敞的,4根支柱穿过建筑延伸出来,指向天际,在地面层形成了一个大型的架空层,这一空间结构源自现代派建筑设计师勒·柯布西耶建筑设计五要素中的"底层架空"理念。

在这里,设计师对空间的把握能力不得不让人惊叹。这个不大的架空空间却是一块休闲的乐土:空间边缘处看似随意放置的各形态巨石,在城市空间内形成了一个巨石公园;方形的游泳池以及外围大面积的"沙滩",似乎突然跳跃到了海边。不得不提的则是攀岩设施。支撑了这个悬空体量的支柱,其实也是一面攀岩墙,虽仍是人造,但它打破了空间的限制——从室内转向了室外,成了真正的户外攀岩设施,给人一种特别的体验。

Design Concept

There is no doubt that this is quite a creative idea. It's not about the architectural form, but offering a new kind of climbing mode – the outdoor climbing between natural rock climbing and indoor climbing. The idea enhances the openness of the building, creating a more direct connection between the building and the city.

Design Feature

Collider Activity Center consists of a huge cube lifted off the ground. The four sides are open. Four great pilotis breaking through the building extend to the sky, creating an extensive open floor. The structure is inspired by the first point of "Le Corbusier's Five Points of Architecture".

Designers' understanding to space is amazing. This relatively small overhead space is a true land of happiness: randomly placed rocks of various shapes at the edge of space create a Boulder Park in the urban space; square swimming pool and extensive "sand beaches" around as if suddenly jump to the seaside. The pilotis supporting the suspending volume act as climbing walls. Even though they are artificial, they manage to break the limitation of spaces from the internal to the external to be the real outdoor climbing facilities offering special experience.

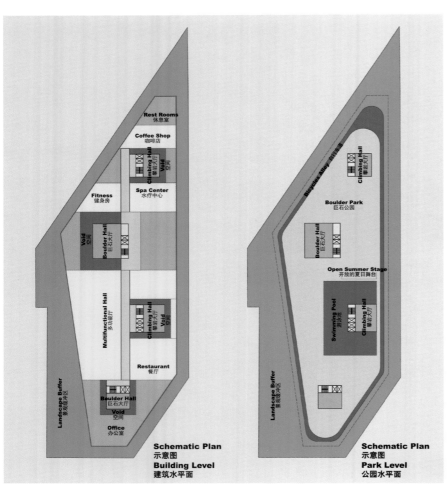
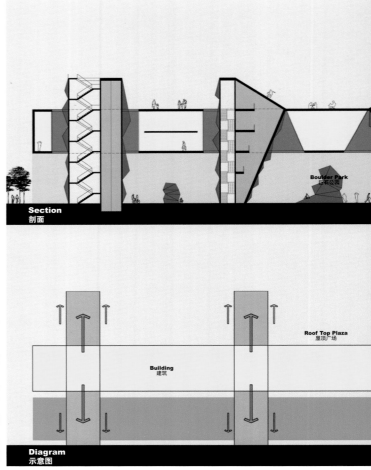

希腊雅典斯帕塔阿提卡公园 Okeanopolis 水族馆
Athens Spata Attica Park Okeanopolis Aquarium

设计单位：Michael Photiadis & Associate Architects
项目地址：希腊雅典
占地面积：16 000 m²
摄影：Ph.Photiadis

Designed by: Michael Photiadis & Associate Architects
Location: Athens, Greece
Site Area: 16,000 m²
Photography: Ph.Photiadis

项目概况

Okeanopolis 水族馆位于斯帕塔阿提卡公园的动物园内，整个项目包括对结构造型、建筑体量以及容纳了 30 个水箱和鱼类生活和展示空间的设计。在这一方案中，有为小型观赏鱼和大型鲨鱼设计的生活空间，也有专属于企鹅、海豹、水獭的水池以及一个海豚观测中心。

设计构思

希腊与海洋有着千丝万缕的联系，正是宽广、柔和而又瞬息万变的海洋，赋予了希腊这个历史悠久的国度惊人的魅力。本方案的设计根植于海洋文化，无论是建筑形态还是室内空间设计，都能依稀发现海洋的踪迹。

设计特色

在形态造型上，设计师避免了对自然形态的直接模仿和复制，而是在参照自然形态的基础上，进行深层次的再创作，最终形成的建筑形态酷似来自海洋中的贝类，又似在海洋中畅游的鱼类、在海上航行的船只或是一湾弧形的水库。

建筑构建在一片水面上，十分切合"水族馆"这一主题。整个体量由多个不同尺寸、相互连接的盒形结构组成，这些基本元素是简单的，但达到的效果却形象而生动：一只贝壳正浮在水面中缓缓爬行。

建筑将游客路径与公园空间分隔开，形成南部水平面上的室内和室外展示区。一条倾斜向下的斜坡延伸至建筑内部，将游客引导至室内空间，似乎将人引入了水平面以下，而室内海蓝色的地板以及展示在透明水箱内的海生物更给人置身于海洋中的错觉，真实而又缥缈。

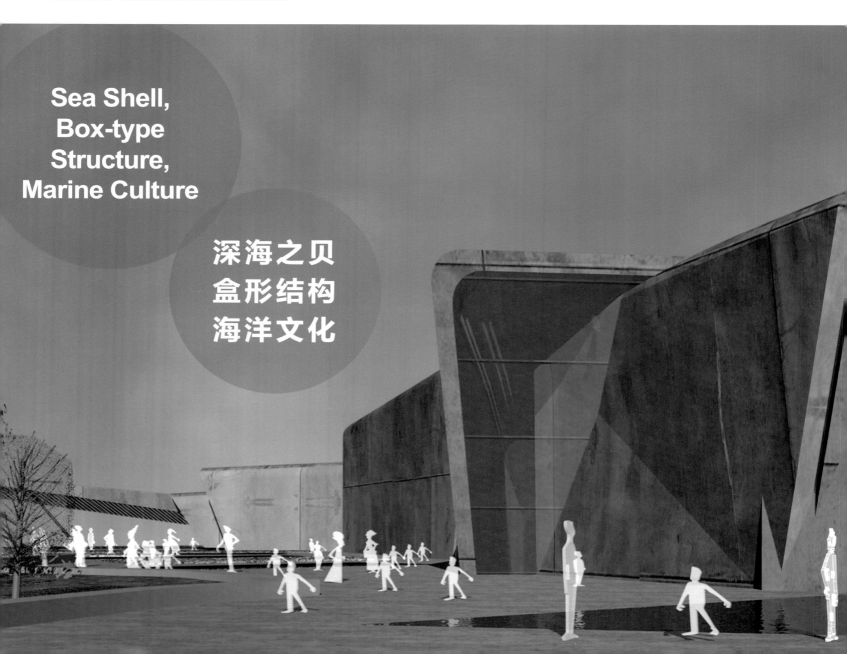

Sea Shell,
Box-type
Structure,
Marine Culture

深海之贝
盒形结构
海洋文化

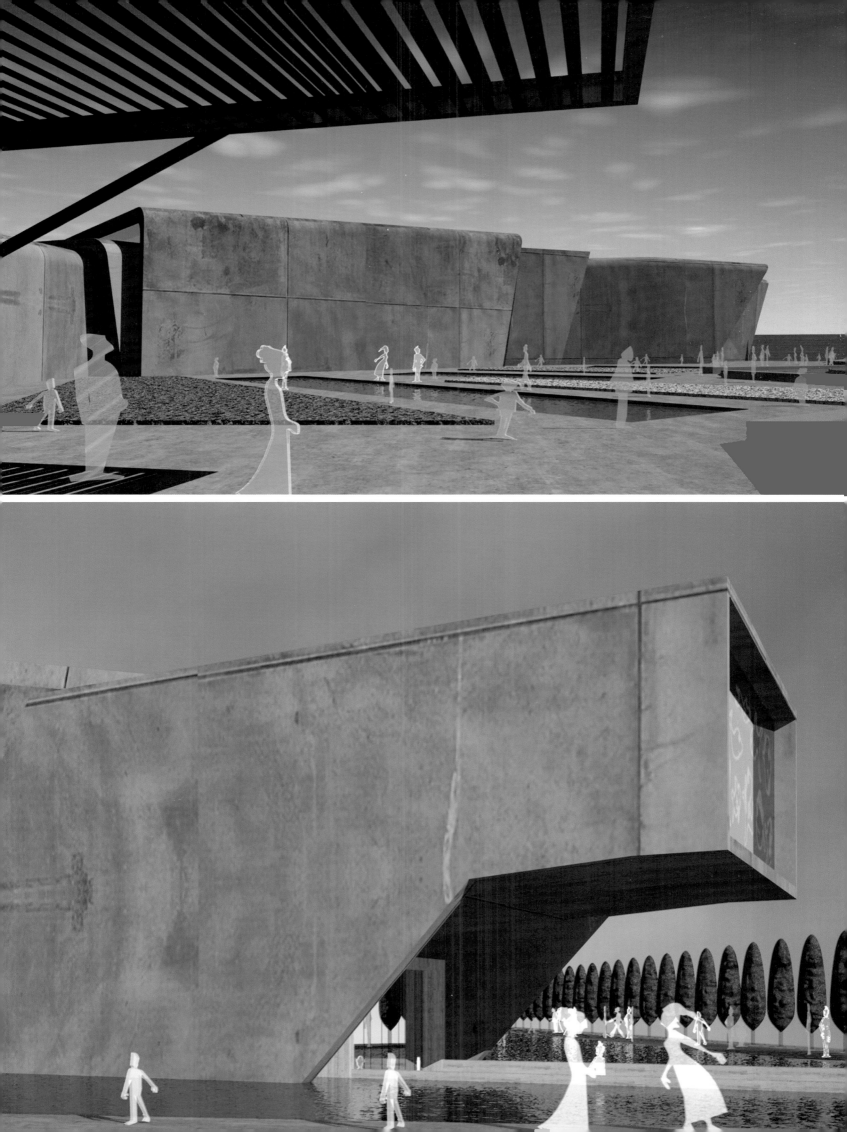

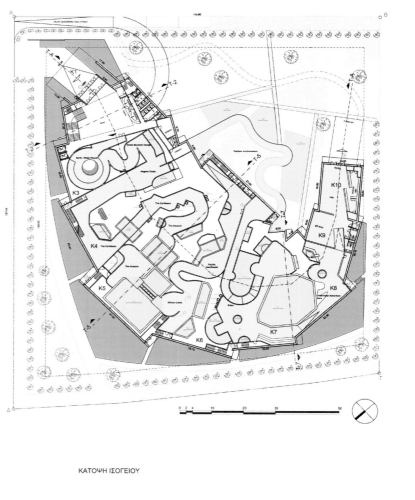

ΚΑΤΟΨΗ ΔΩΜΑΤΟΣ ΠΕΡΙΒΑΛΛΟΝΤΟΣ

ΚΑΤΟΨΗ ΙΣΟΓΕΙΟΥ

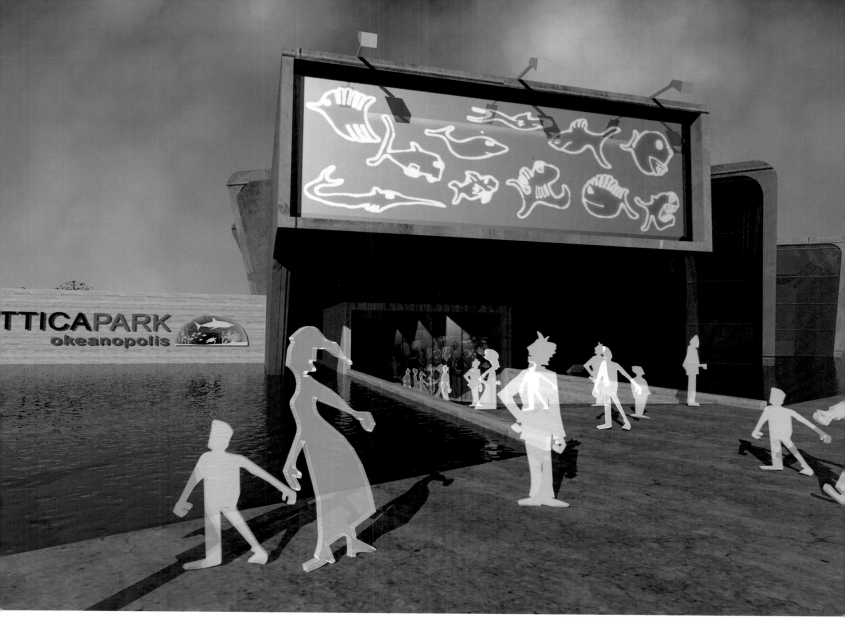

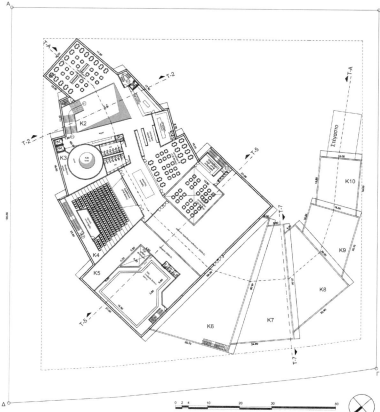

ΚΑΤΟΨΗ ΟΡΟΦΟΥ

Profile
Okeanopolis Aquarium is located at the Attic Park Zoo. The whole project includes the structural shape, volume, outer and inner spaces for the 30 tanks and fish spaces. There are living spaces for Aegean fish and monster sharks, separate lagoons for penguins, seals and otters and a lively dolphin viewing center.

Design Concept
Greece is inextricably linked with the ocean. It is the extensive, gentle, ever-changing ocean that endows the history of Greece with stunning charm. This scheme is rooted in marine culture – no matter in the architectural form or the interior space design, ocean elements can be found.

Design Feature
The final image purposely avoids the imitation or copy of explicit natural shapes but relates to sea water shells – fish creatures, shipwrecks or reservoirs based on reference of natural forms and further re-creation.

The building is constructed above water, which fits in with the theme of "Aquarium". The volume is broken into articulated boxes of various sizes. These simple elements manage to achieve a vivid and lively effect: a gliding shell floating or emerging from water.

The structure's location divides the guest trail from the park to the interior and exterior displays in the south level. Through an entry of a downward ramp guests feel guided below water level. The internal ocean blue floors and marine creatures in transparent tanks create an illusion that people are staying in the ocean.

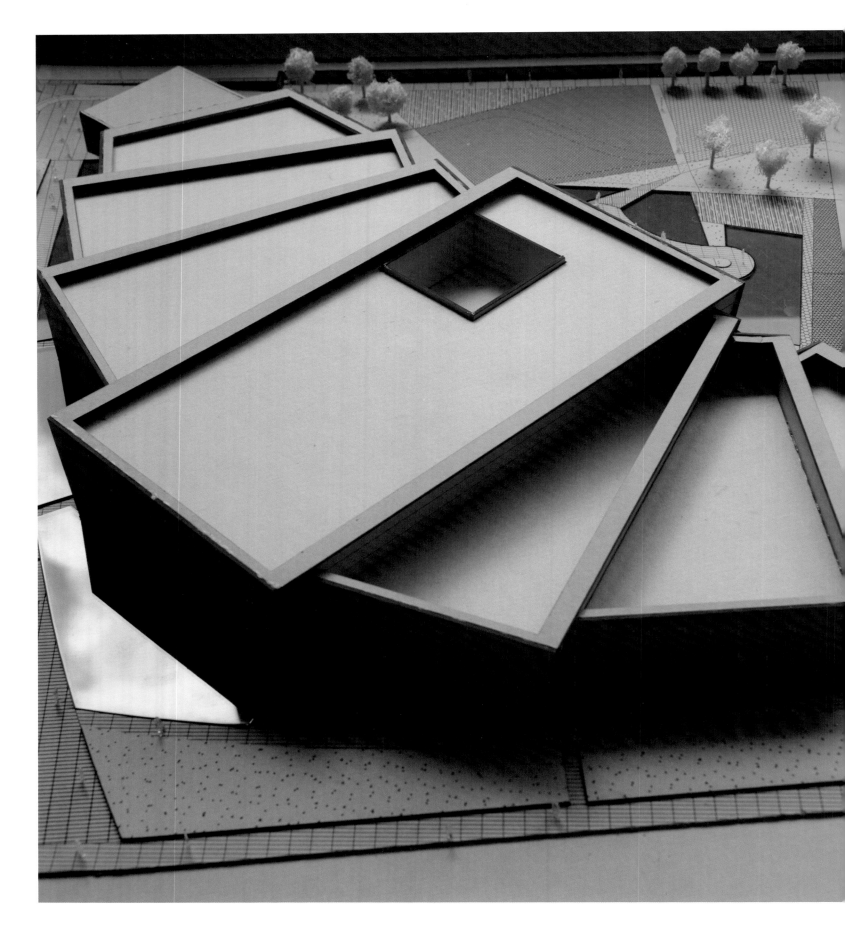

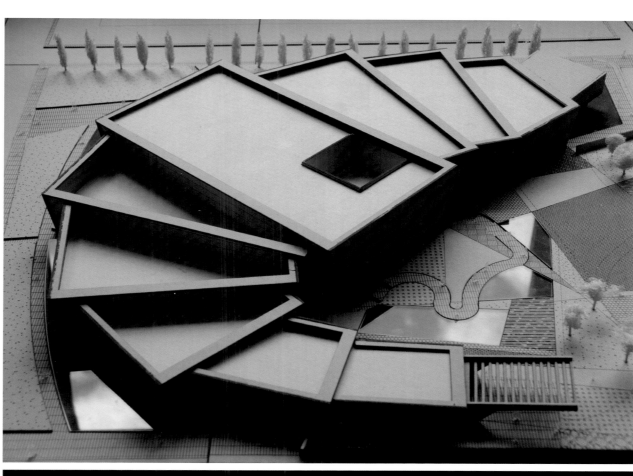

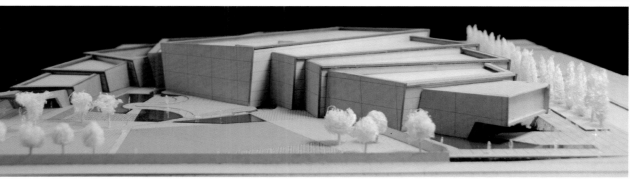

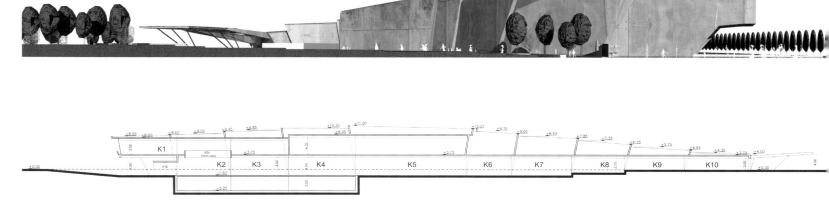

"Lotus Leaf", "Water Lily", Free & Fluid

"荷叶" "睡莲" 鱼戏莲间

浙江杭州西溪休闲中心
Xixi Leisure Center

设计单位：DnA_Design and Architecture
开发商：杭州西溪国家湿地公园三期工程有限公司
项目地址：中国浙江省杭州市
建筑面积：6 300 m²

Designed by: DnA_Design and Architecture
Developer: Hangzhou Xixi National Wetland Park Phase III Engineering Co., Ltd.
Location: Hangzhou, Zhejiang, China
Floor Area: 6, 300 m²

项目概况

西溪休闲中心是坐落在西溪国家湿地公园的12个艺术和文化建筑群的单体建筑之一，这些建筑群体的建设将丰富杭州的文化底蕴，推动当地旅游业的发展。

设计构思

鱼戏荷叶间的自由畅快，不经意间就勾起了内心的向往和憧憬。长期生活在现代化环境中的人们是否能像千百年前的先祖那样如轻灵的鸿毛、敏捷的祥龙一般在充满微风和清香的自然环境中自由逗留？设计师从亭亭的荷叶获取灵感，设想了一个连续、开敞、自然的空间，使人们可在其中自由穿梭，获得一份独特的体验。

设计特色

这个形如亭亭荷叶的单体建筑，形成了多个错落有致、环环相扣的休闲空间。休闲中心延续了原有湿地的路径和形态，营造出由地面上的休闲功能空间和下沉的活动池塘组成的连续、开敞的空间形态。位于二层的特色水疗室成为相对独立的空间，好似漂浮在水上的睡莲。

这种形似荷叶的空间组织和连接方式使人们从户外的步行通道经由中间各层的连接通道，穿行于层层的"荷叶"间，宛若鱼儿嬉戏在荷叶间般自由畅快。及至到达屋顶上的小池塘，才揭开这片湿地的面纱，在此俯瞰建筑全貌，也更惊叹设计师的匠心和巧思。

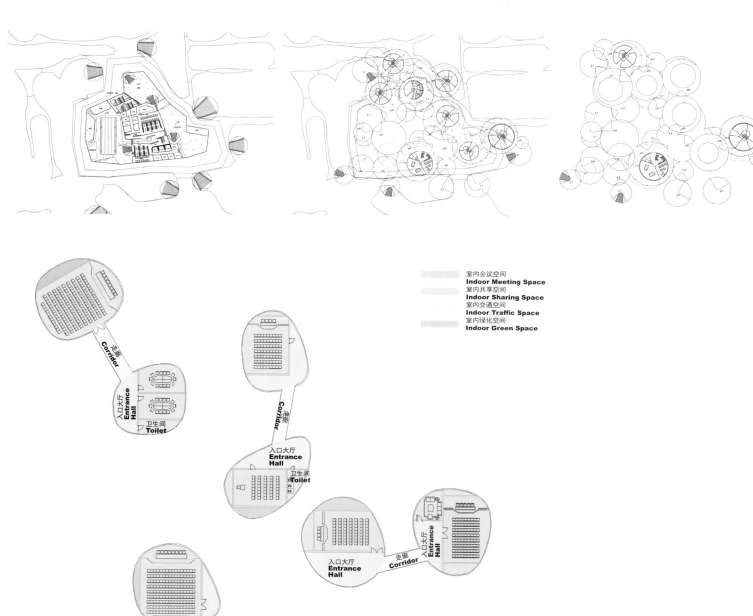

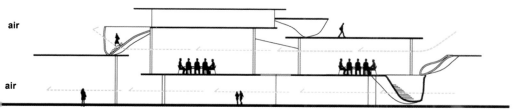

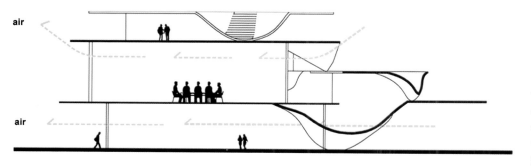

Profile

Xixi Leisure Center is one of the 12 buildings located in an art and culture compound in Xixi National Wetland Park. These buildings shall enrich Hangzhou's cultural deposits and promote the development of local tourism.

Design Concept

A pause on these leaves might evoke inner desire and longing. We human beings, after longtime living in a modern environment, are able to linger amongst a natural environment filled by breeze and aroma, light as a feather, alert as a dragon? Inspired by graceful lotus leaves, designers envisaged a continuous, open natural space that people shall get a special experience when passing by.

Design Feature

The lotus leaf-shaped building forms multiple staggered & interrelated leisure spaces. The leisure center will house leisure function as open and continuous circulation on ground level and sunken into activity pools imitating paths and ponds of wetland topography. On the second level, specialized SPA rooms become rather individual spaces, like water lilies floating on water. The format and organization of leaves leads an outdoor promenade from the paths up to roof terraces discovering small water lily ponds, eventually unveiling the wetland to a panoramic view from above. The originality and ingenuity are quite surprising.

荷兰乌特勒支 Belle van Zuylen 大楼
Belle van Zuylen Tower

设计单位：de Architekten Cie.
开发商：Burgfonds, Zaltbommel
项目地址：荷兰乌特勒支
设计团队：Pi de Bruijn　Branimir Medić
　　　　　J. Lee　　　M. Campschroer
　　　　　T. Deutinger

Designed by: de Architekten Cie.
Client: Burgfonds, Zaltbommel
Location: Utrecht, Netherlands
Design Team: Pi de Bruijn, Branimir Medić,
J. Lee, M. Campschroer, T. Deutinger

精致
"花瓶"
垂直连接
空中景观

Exquisite
Vase Shape,
Vertical Connection,
Air Landscape

设计构思

设计师设想在乌特勒支市构建一栋超高层的建筑，通过合理的布局实现建筑功能在水平布局与垂直布局之间的平衡，从而为该市的城市规划引入一种新的发展模式。

设计特色

Belle van Zuylen 大楼高 262m，宛如一尊精致的花瓶，又好似记载时光的沙漏，优雅、卓尔不群的形态使其具备了惊人的视觉震撼力，新颖的设计概念则引领它走在时代的前沿。项目的基本功能区以住宅区和办公区为主，同时还涵盖了购物、娱乐、咖啡厅、餐厅以及文化活动区，完整的功能设置使建筑基于"整合营造和谐的整体环境所需的功能区"的理念得以实现。

为维持当地原有的景观环境，城市功能区主要以垂直方式连接。同时，露台花园、庭院花园及水景将垂直分布在不同的楼层内，贯穿整个建筑，从底层一直扩展至顶层，将公共区和生态功能区自然地整合到建筑中。

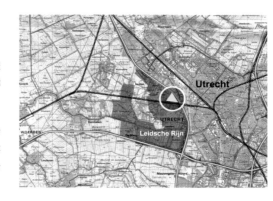

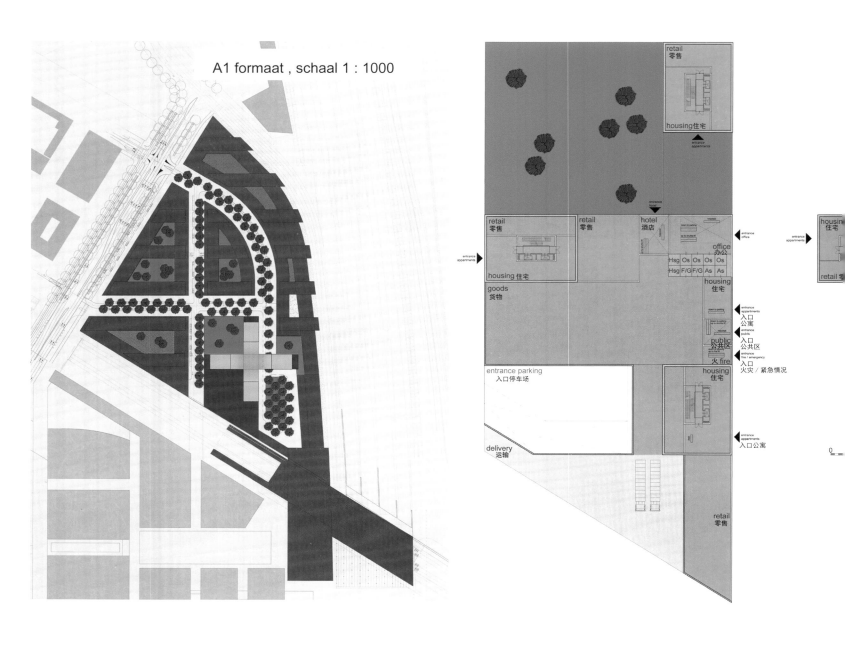

A1 formaat , schaal 1 : 1000

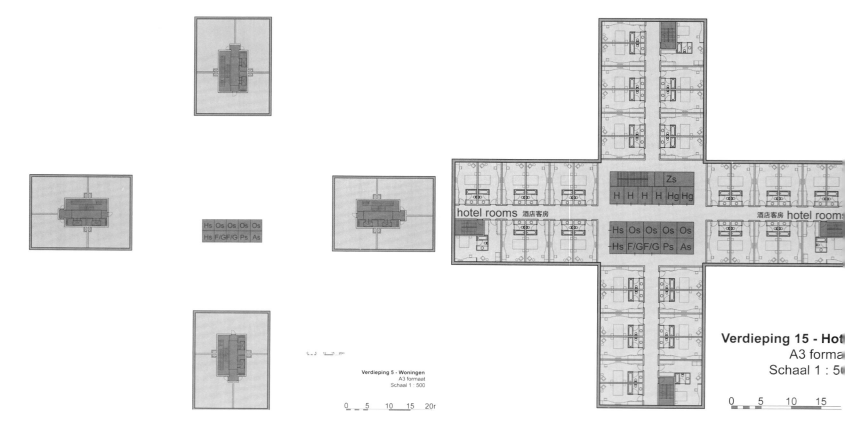

Verdieping 5 - Woningen
A3 formaat
Schaal 1 : 500

Verdieping 15 - Hotel
A3 formaat
Schaal 1 : 500

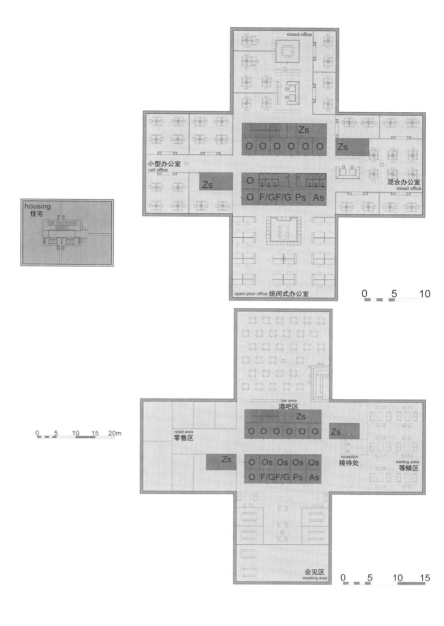
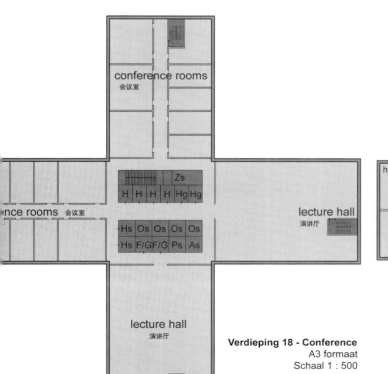

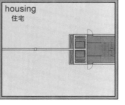
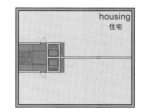

Verdieping 18 - Conference
A3 formaat
Schaal 1 : 500

Verdieping 65 - Woningen
A3 formaat
Schaal 1 : 500

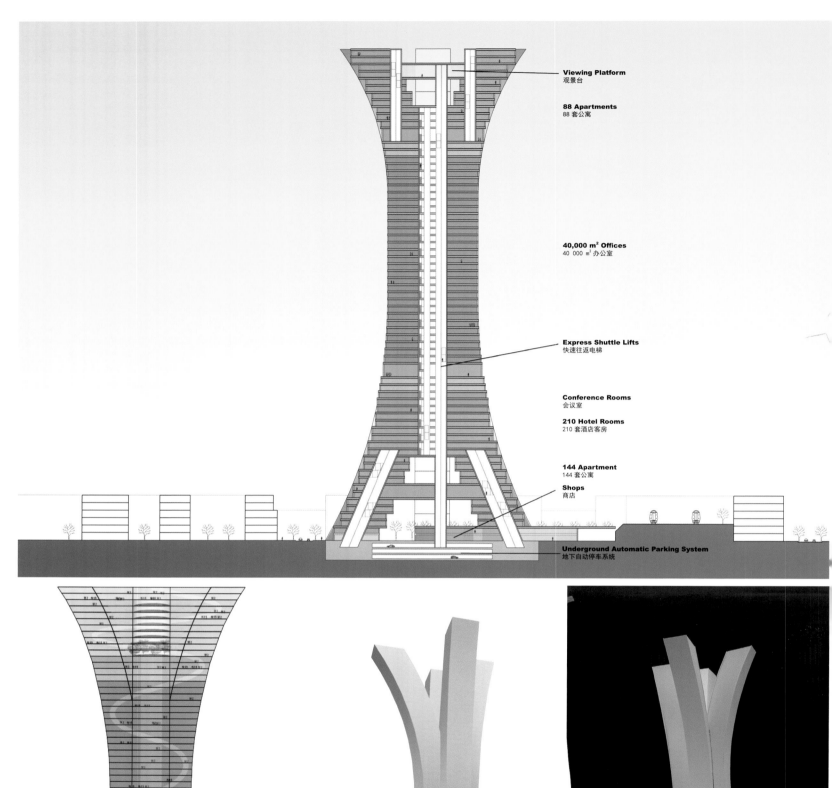

Viewing Platform
观景台

88 Apartments
88 套公寓

40,000 m² Offices
40 000 m² 办公室

Express Shuttle Lifts
快速往返电梯

Conference Rooms
会议室

210 Hotel Rooms
210 套酒店客房

144 Apartment
144 套公寓

Shops
商店

Underground Automatic Parking System
地下自动停车系统

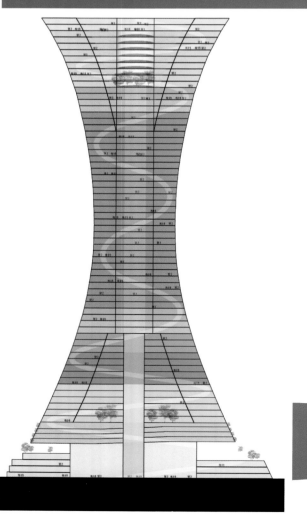

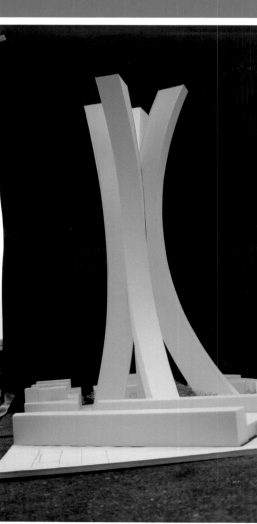

Design Concept

Designers have envisaged building a super high rise in Utrecht City. It provides an opportunity to introduce a new grammar in urban planning, in which the horizontal and vertical communicate and dovetail with one another in a wholly logical manner.

Design Feature

With tremendous height of 262 meters, the Belle van Zuylen stands for grace and elegance, imposing stunning visual impact. Its novel design ideas lead its way ahead of the times. The primary functions are living and working, but there is also room for shopping, recreation, cafés and restaurants, and cultural activities. The core of the concept on which the Belle van Zuylen is based lies in the combination of functions that is necessary to create a congenial and complete environment.

To keep much of the local existing landscape untouched, the urban functions are vertically interlinked in a logical manner. Terrace gardens, courtyard gardens and water features will be realized at various levels throughout the building, from the ground floor to the very summit, so as to naturally integrate public areas and ecological functional areas to the tower.

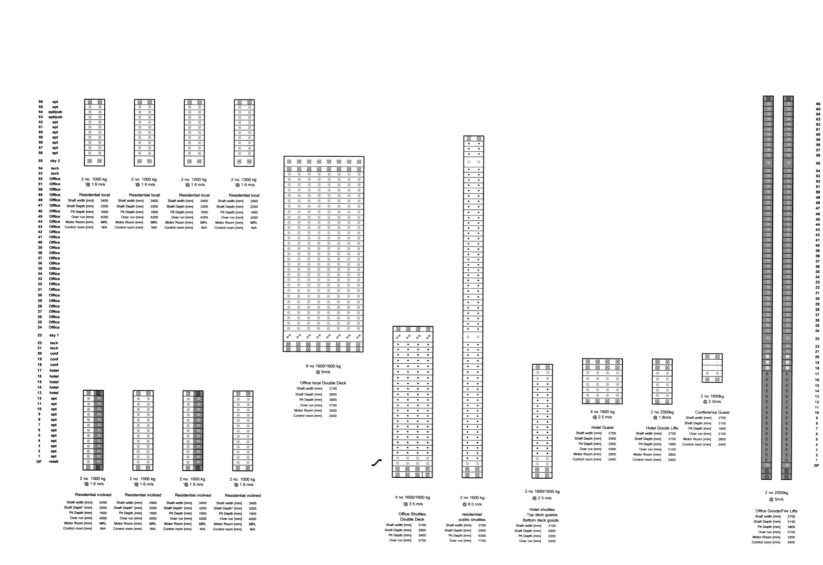

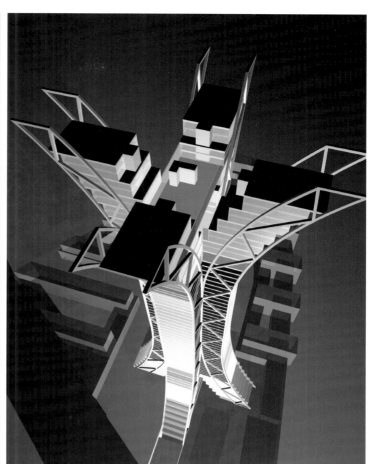
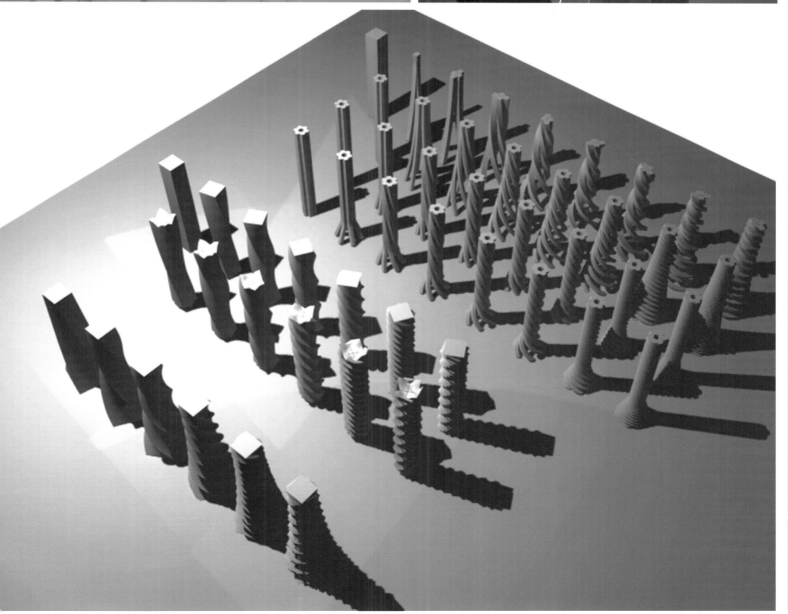

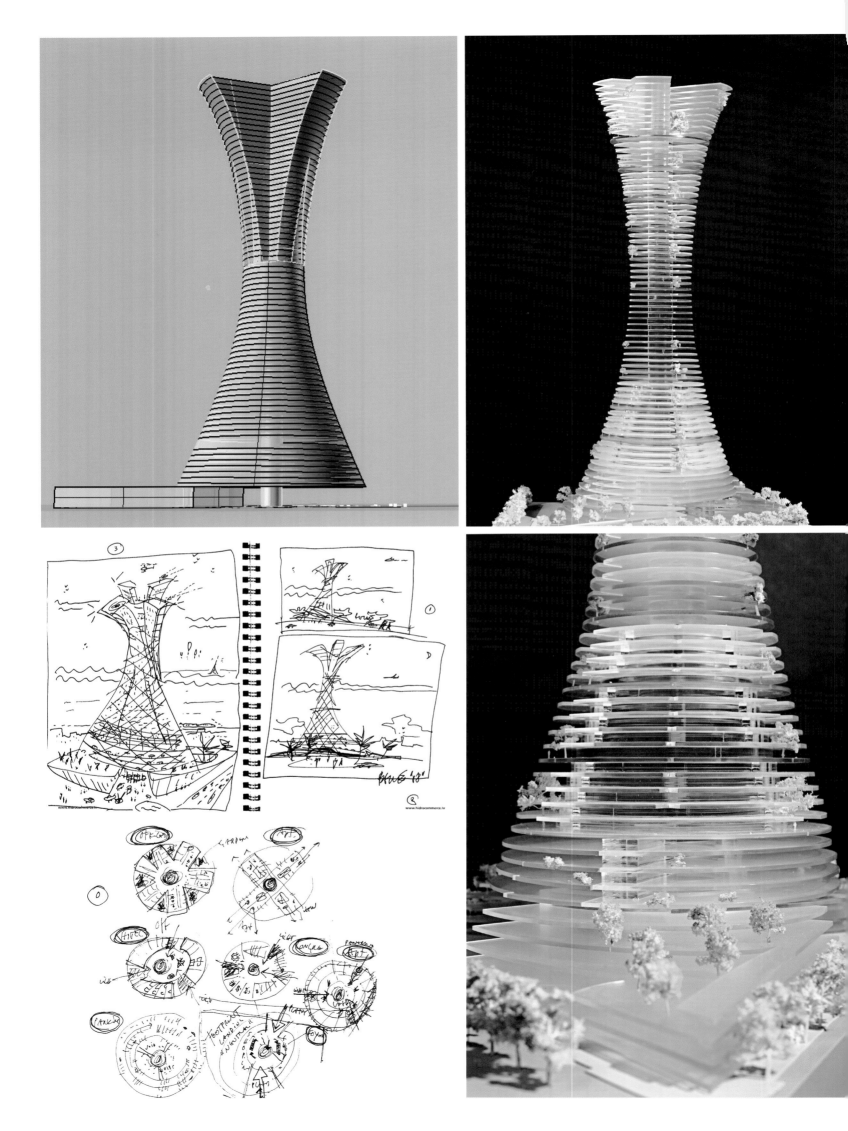

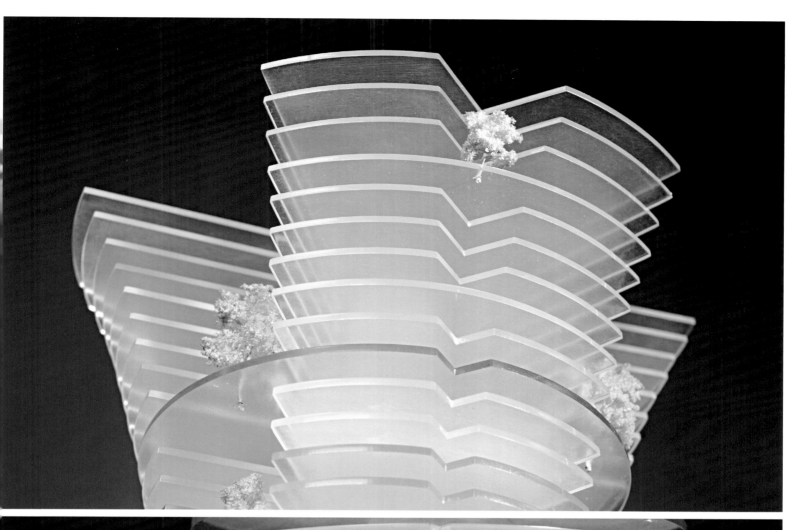

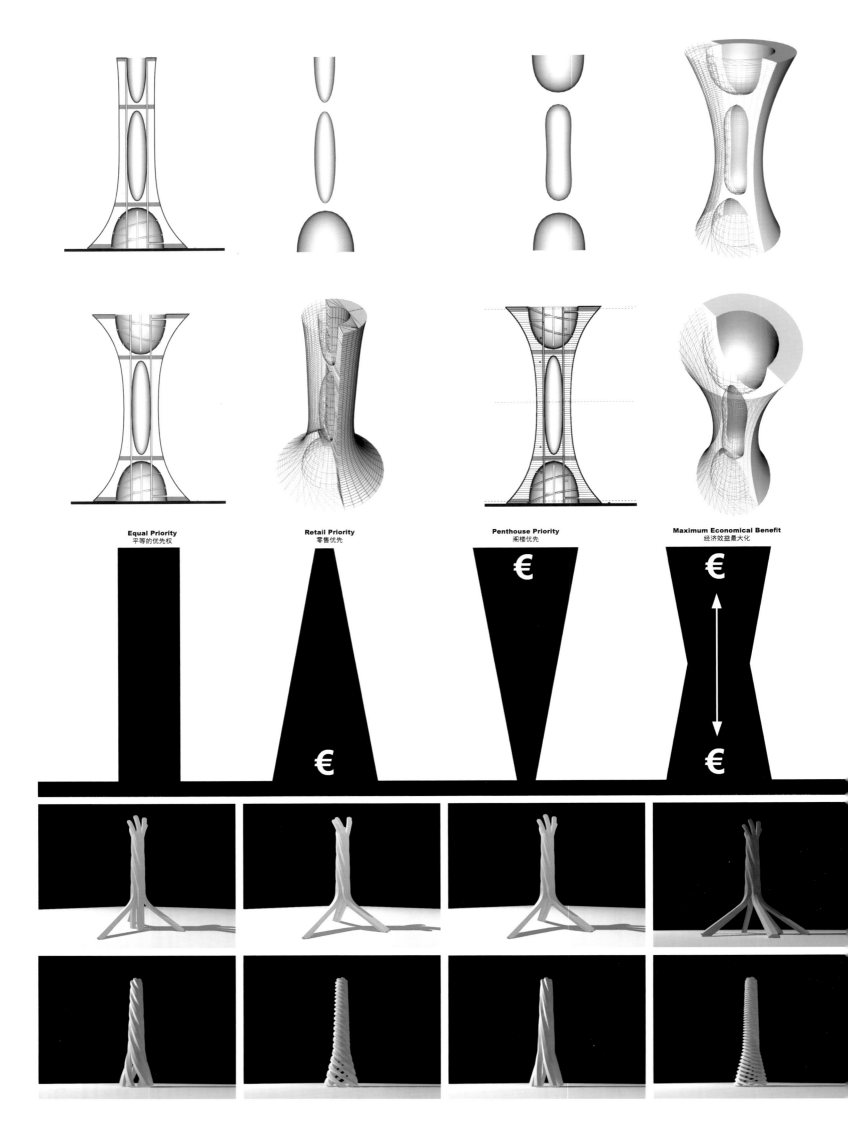

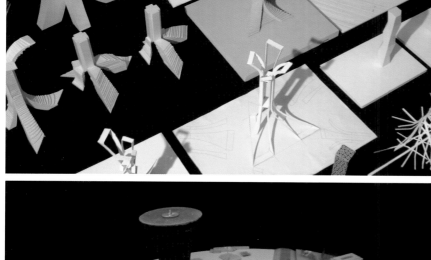
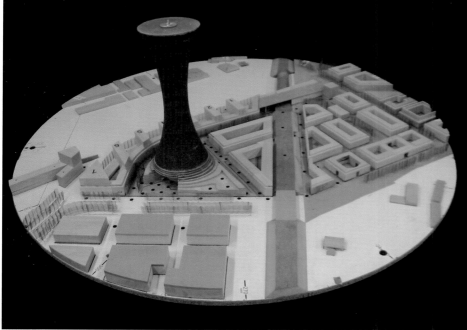
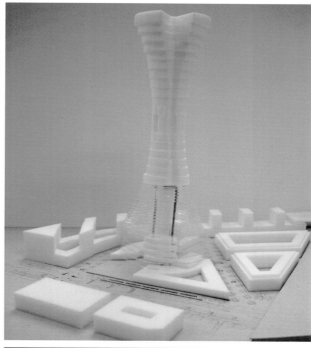
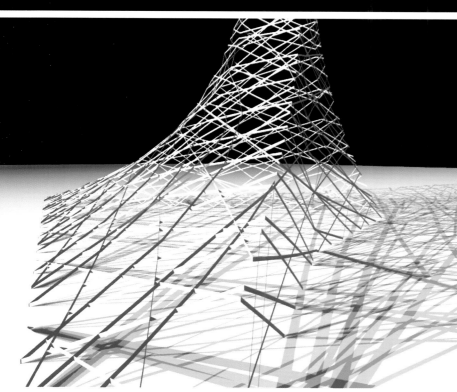
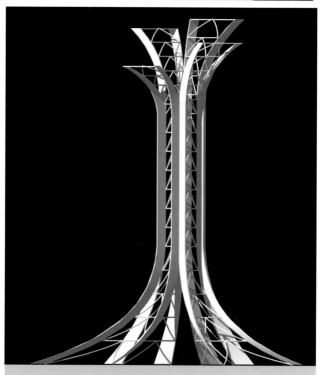

丹麦 Køge 大学医院
Køge University Hospital

设计单位：C. F. Møller Architects
项目地址：丹麦 Køge
总建筑面积：177 000 m²（其中新建部分 130 000 m²）

Designed by: C. F. Møller
Location: Køge, Denmark
Gross Floor Area: 177,000 m² (Newly Constructed: 130,000 m²)

技术分散设计
智能化

Decentralized Technical Installations, Intelligent Design

项目概况

本方案是对 Køge 大学医院进行扩建的方案，这一方案无论是在结构上还是功能上都十分有远见。设计以已有医院的品质和发展潜力为基础，展现了一个结构清晰、布局紧凑、绿色且富有吸引力的医院综合体在可持续、功能和技术方面的愿景。

设计特色

按照常规，电力、HVAS、通风、防火等技术装置通常都会集中设计，然而在这一方案中，设计师却反其道而行之，将它们分散开来，设置在每一个房间周围。看似简单的改动，却能实现对建筑空间的最优化利用，同时也提高了空间的灵活度，使每一个房间可在未来的规划中依据相关技术设施的需求进行灵活调整。

创新的后勤方案还体现在内部后勤系统自动化上。项目将采用移动机器人与自导车辆系统送餐、递送衣物和药品、收拾餐具等，另外还将采用气压输送管系统，这些智能化设计，极大地解放了人力资源，使患者可以得到更好的照顾。

建筑设计

原有的住院大楼将被拆除，主大厅则将作为新医院的主要通道而保留下来，并会进一步扩建，新增庭院花园及屋顶天窗。从主大厅有四条垂直的交通路线通向中央广场，每一条路径都连接着一个庭院花园。考虑到建筑的紧凑性，新建筑的走廊将适当减少，以组织更便捷的人行流线，缩短员工的步行距离。同时，紧凑的功能布局也为医院在未来进一步扩建预留了空间。

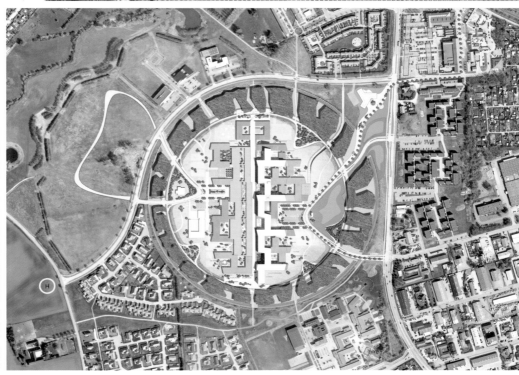

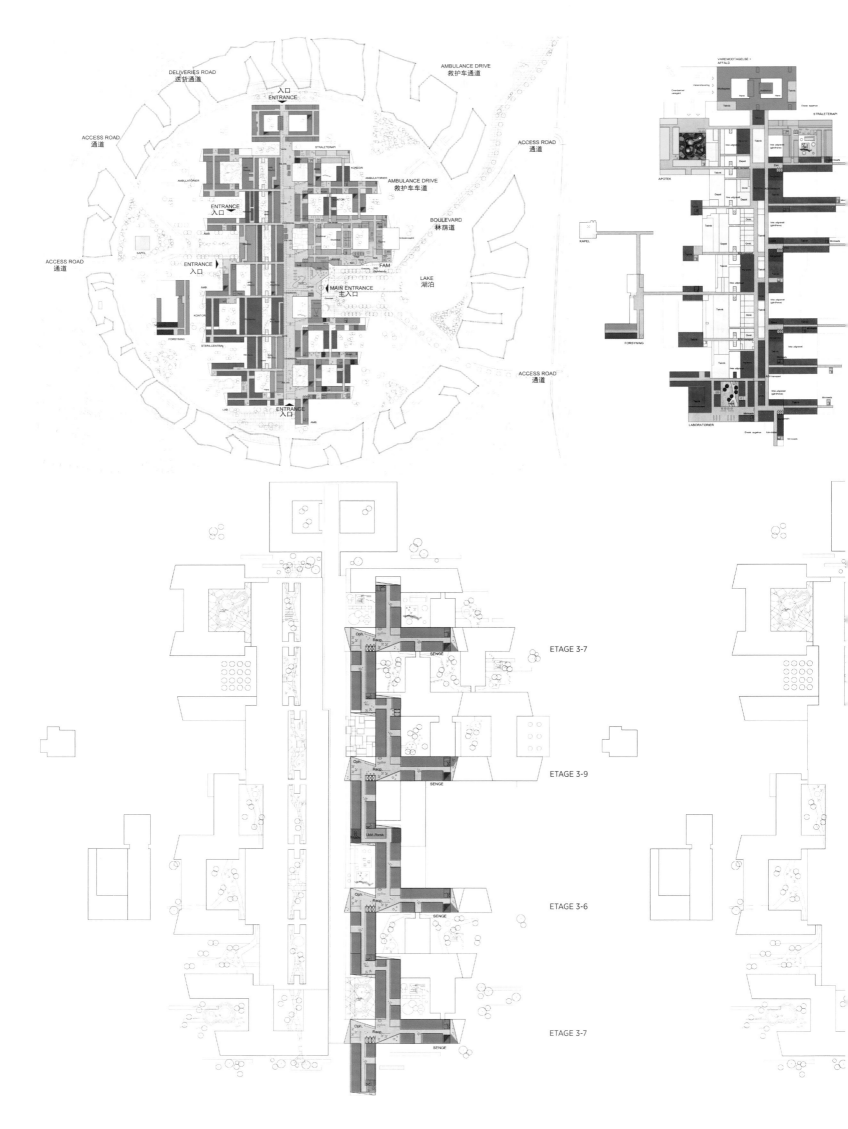

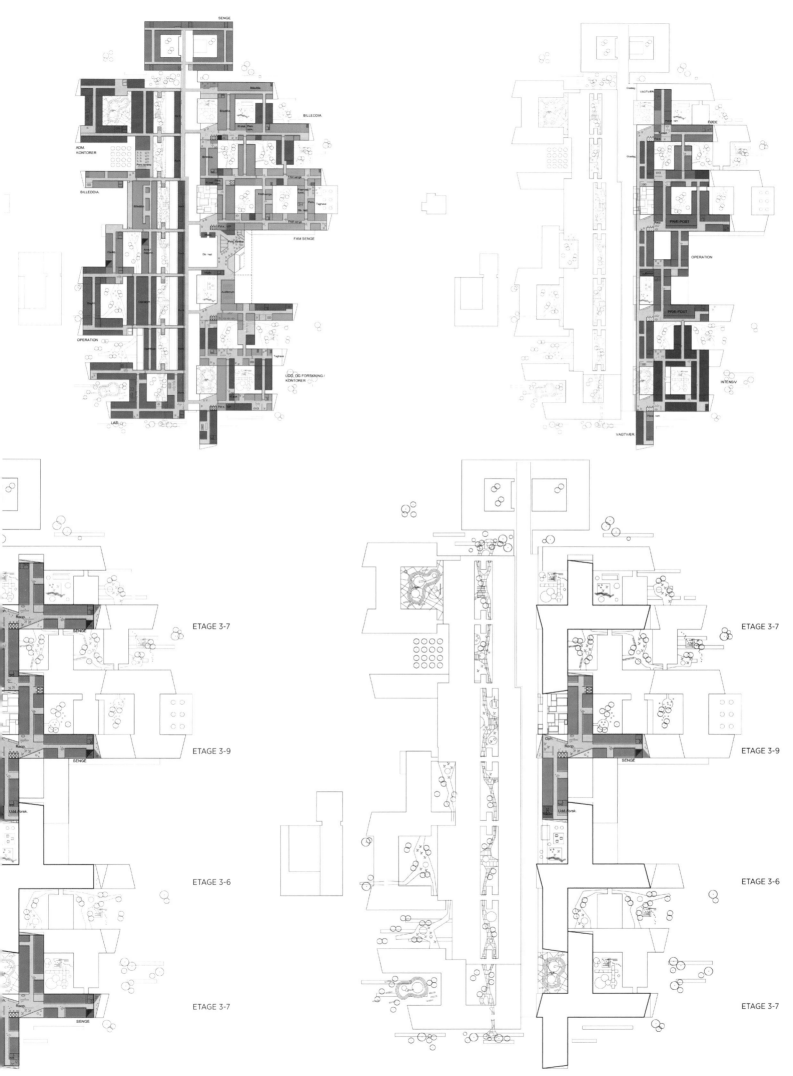

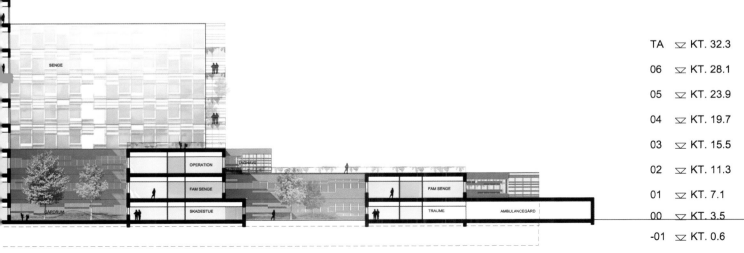

TA	▽ KT. 32.3
06	▽ KT. 28.1
05	▽ KT. 23.9
04	▽ KT. 19.7
03	▽ KT. 15.5
02	▽ KT. 11.3
01	▽ KT. 7.1
00	▽ KT. 3.5
-01	▽ KT. 0.6

Tværsnit 1:500

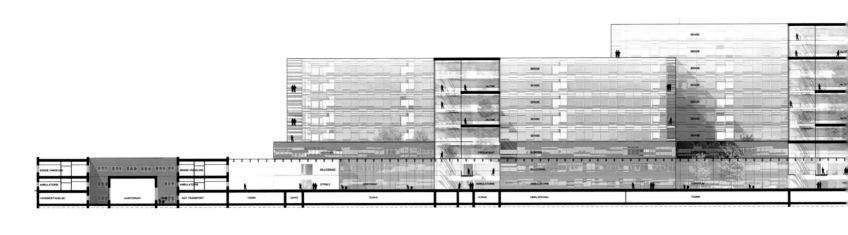

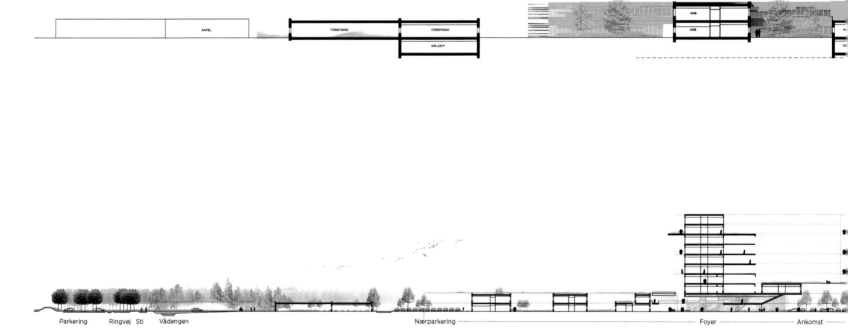

Parkering　　Ringvej　Sti　Vådengen　　　　　　　　　　Nærparkering　　　　　　　　　Foyer　　　　　Ankomst

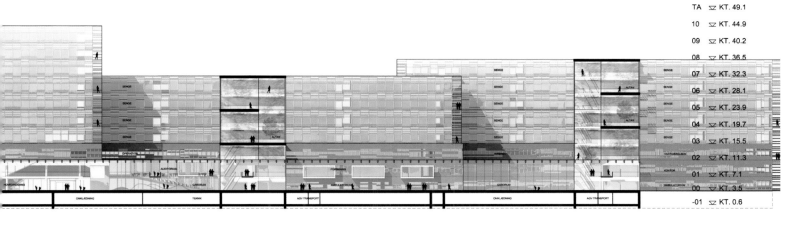

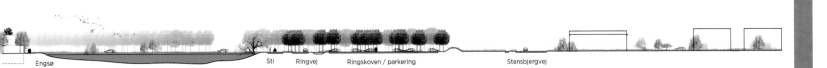

Profile

The project is an expansion of the existing Køge Hospital. It is a visionary project in terms of both architecture and functionality. The project is based on the existing hospital's qualities and potential to present a cohesive sustainable, architectural, functional and technical vision of a clear, compact, green and inviting hospital complex.

Design Feature

Technical installations such as electricity, HVAS, ventilation, fire prevention measures etc. are traditionally centrally located in a hospital. But in this case, they will be decentralized and located adjacent to each room. This will make optimum use of the available building space, improve space flexibility and adjust the functions and related technical supplies of each room.

The innovative logistics solution also reflects on the internal logistics systems automation. Mobile robots, or AGVs are applied to transport meals, linen, medicine, used tableware, etc. A pneumatic tube post system is also used. The intelligent design will release staff resources for the care and treatment of patients.

Architectural Design

The existing inpatient buildings will be demolished in order to retain the present main hall as the new hospital's main thoroughfare. The main hall will be expanded and new courtyard gardens and roof lighting will be added. From here, four vertical transport routes will lead to central squares, each with their own related courtyard garden area. Due to the compactness of the building, there will be fewer corridors in the new hospital, and shorter distances for staff to walk. Meanwhile, the compact function layout also reserves spaces for future expansion.

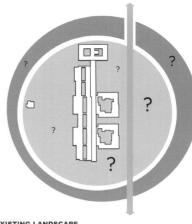

ANALYSIS OF EXISTING LANDSCAPE
原有景观分析
The circular forest is a significant and unique large-scale feature. Its potential as a spatial experience and transition zone is however not fully resolved. Despite good connections to the local surroundings, the open clearing is mostly a transit space.
环形森林是该项目一个显著的独特特征。它作为一个空间体验和过渡区的潜能没有完全被瓦解。这一个作为过渡空间的开放空地，与当地环境有良好的联系。

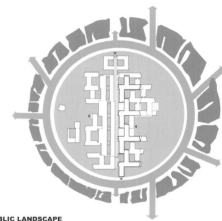

NEW PUBLIC LANDSCAPE
新公共景观
A new landscape feature, the wetland, is introduced. The wetlands are situated along the inside of the circular forest, and animate the inner landscape around the hospital. New paths along the wetlands connect the re-designed inner landscape with the immediate surroundings, and improve the accessibility of the hospital grounds.
引入一个新的景观元素——湿地。湿地沿环形森林内部分布，激活了医院周围的内部景观。沿湿地的新道路将重新设计的内部景观与周围环境相连接，提升了医院场地的可达性。

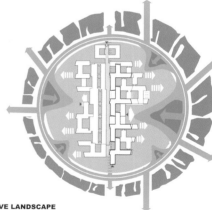

NEW ACTIVE LANDSCAPE
新的活跃景观
Pockets of landscaped parks and activity areas along the wetlands create attractive green spaces circling the new hospital, accessible to both the neighbouring community and the users of the hospital. The open clearing is transformed from a passive transit space into a new public attraction.
沿湿地的景观公园和活动区营造了围绕新医院周围的富有吸引力的绿色空间，周围社区及医院的患者都可以利用这些绿色空间。这一个开放的空地从一个被动的过渡空间转变成了一个新的公共吸引点。

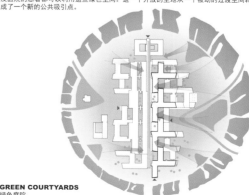

GREEN COURTYARDS
绿色庭院
The green landscape is extended throughout the internal street and interlaced with the hospital's courtyards.
绿色景观贯穿整个内街，与医院的庭院交织在一起。

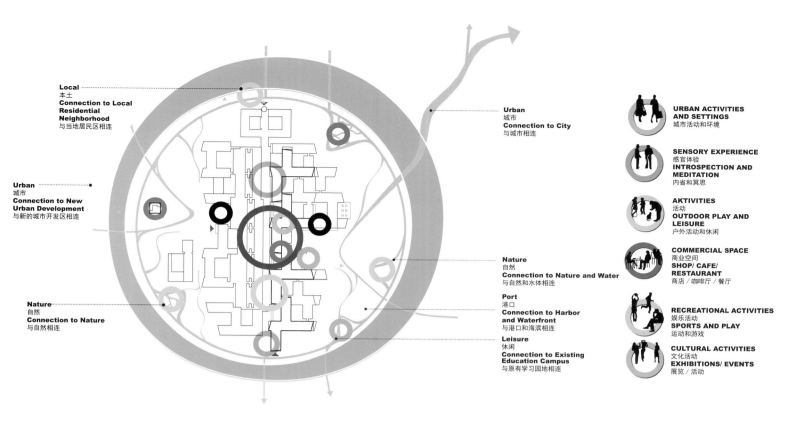

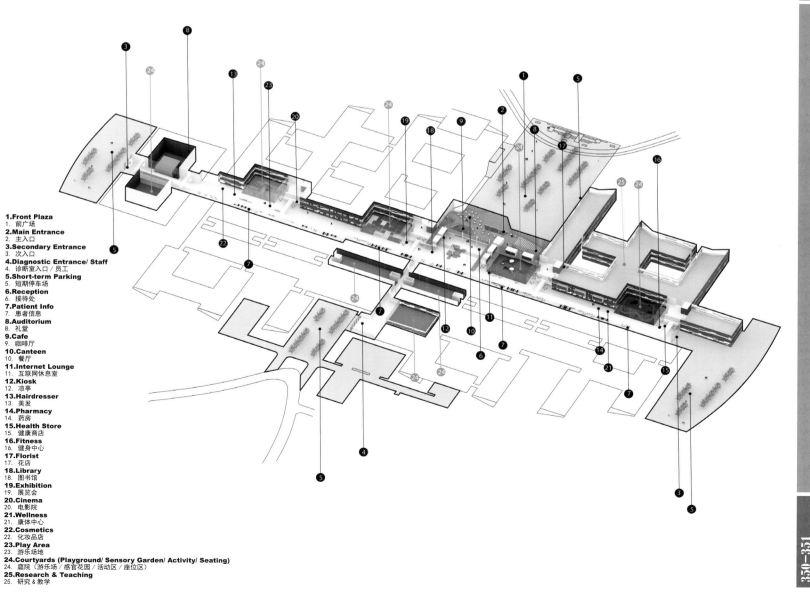

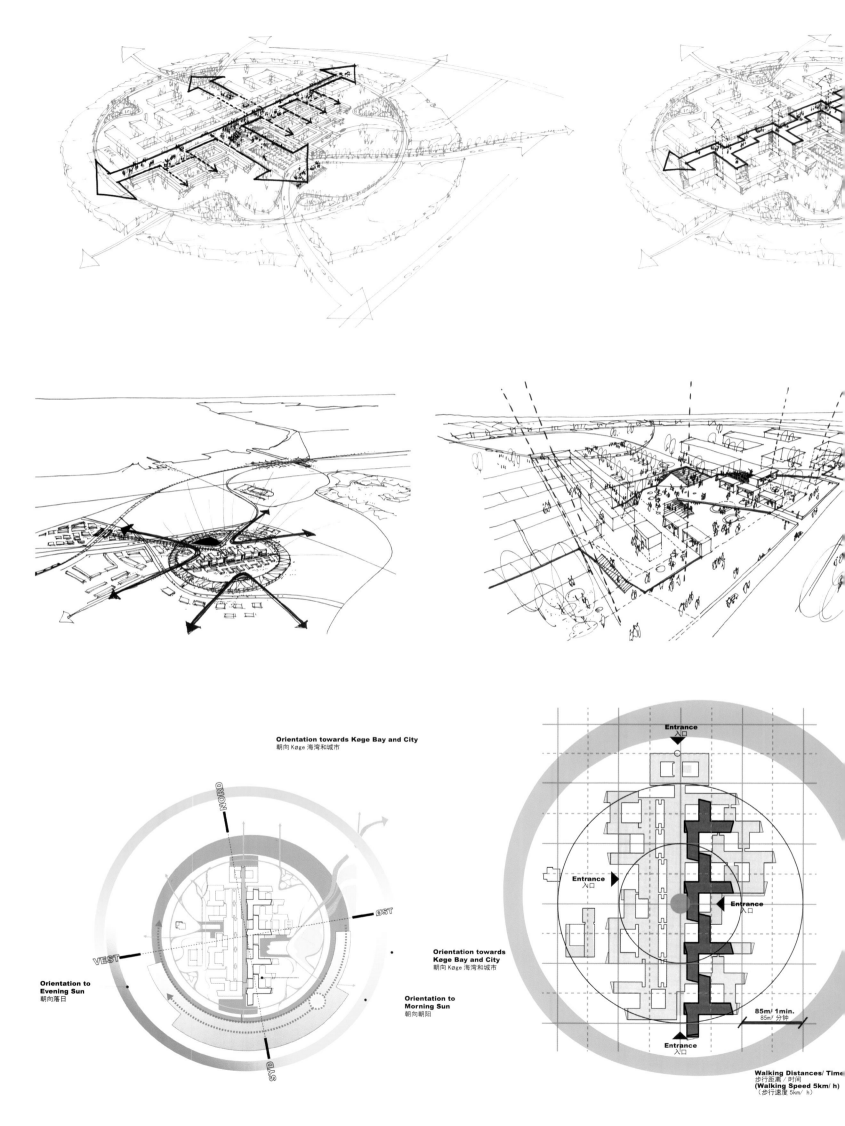

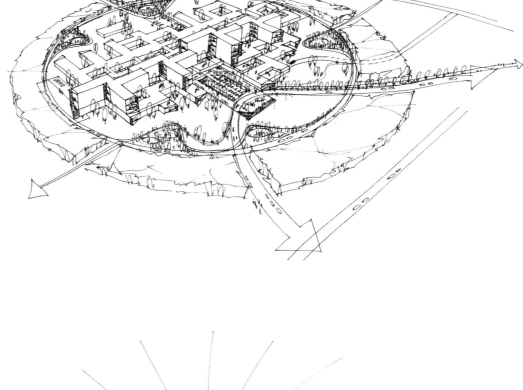

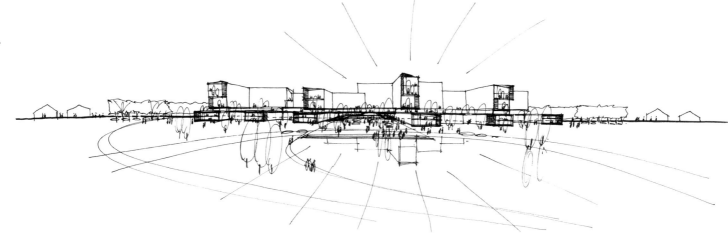

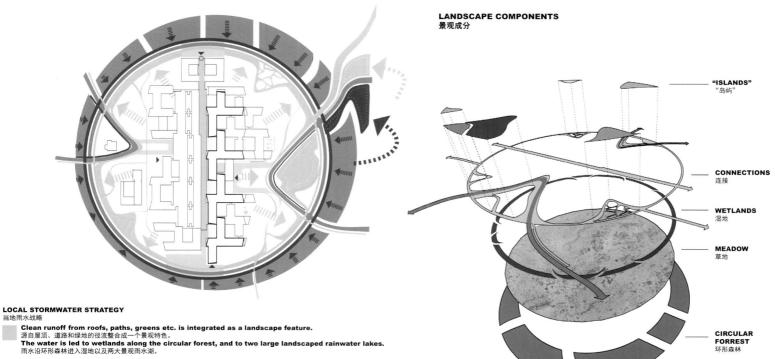

LANDSCAPE COMPONENTS
景观成分

"ISLANDS"
"岛屿"

CONNECTIONS
连接

WETLANDS
湿地

MEADOW
草地

CIRCULAR FORREST
环形森林

LOCAL STORMWATER STRATEGY
当地雨水战略

Clean runoff from roofs, paths, greens etc. is integrated as a landscape feature.
源自屋顶、道路和绿地的径流整合成一个景观特色。
The water is led to wetlands along the circular forest, and to two large landscaped rainwater lakes.
雨水沿环形森林进入湿地以及两大景观雨水湖。

Runoff from roads and parking areas is collected in buffer volumes along the ring road and filtered in a separate pool, before it is canalized to the north-east rainwater lake.
沿环形道路的缓冲区体量收集道路和停车区的径流，在一个独立的水池中过滤以后汇入东北方的雨水湖。

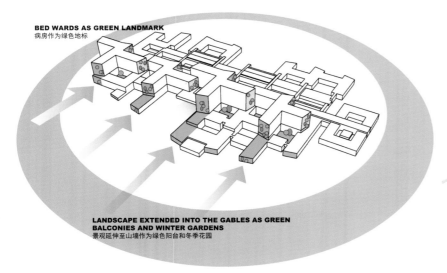

BED WARDS AS GREEN LANDMARK
病房作为绿色地标

LANDSCAPE EXTENDED INTO THE GABLES AS GREEN BALCONIES AND WINTER GARDENS
景观延伸至山墙作为绿色阳台和冬季花园

GREEN LANDMARK
绿色地标

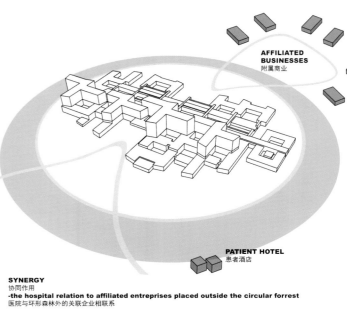

AFFILIATED BUSINESSES
附属商业

PATIENT HOTEL
患者酒店

SYNERGY
协同作用
- the hospital relation to affiliated entreprises placed outside the circular forrest
医院与环形森林外的关联企业相联系

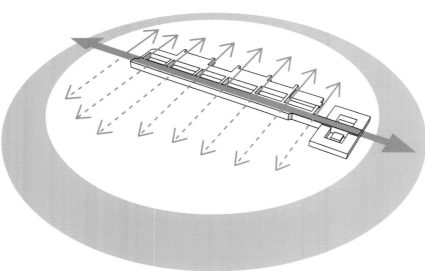

VISION
愿景
to create a logistically efficient, seamless and eventful hospital of international standards
打造一个高效、无缝结合的国际标准医院

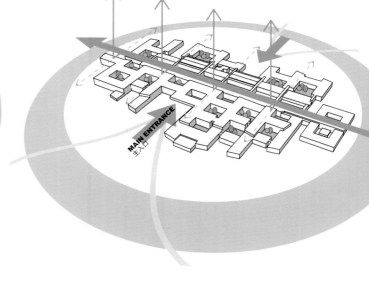

MAIN ENTRANCE
主入口

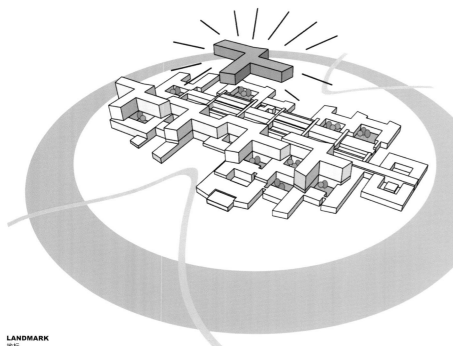

LANDMARK
地标
Bed ward building as green landmark
病房建筑作为绿色地标

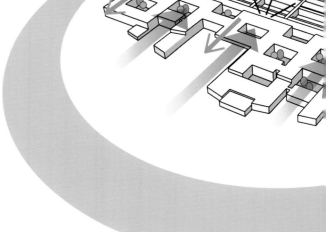

DISPLACEMENT
位移
opens the structure to the outside and improves daylight and views in the internal street
将建筑结构向室外打开，提升内街的采光和视野

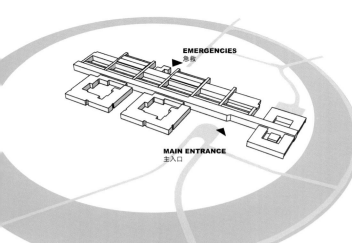

EXISTING HOSPITAL
现有医院

VISION
愿景
exiting flow of the Køge Hospital
Køge 医院的已有流线

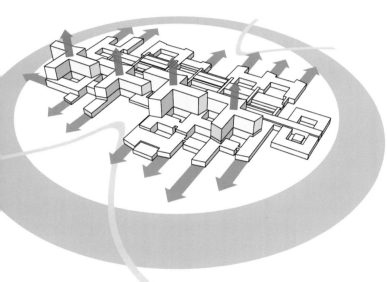

FUTURE EXTENSION POTENTIAL
未来扩建潜力

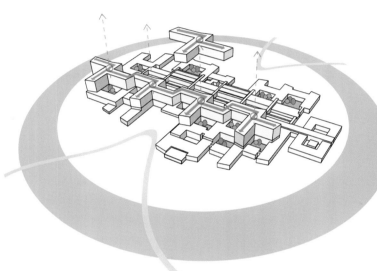

BED WARDS
病房
Easy orientation in the wards from the 4 elevator - cores
从四个电梯很容易对病房定位

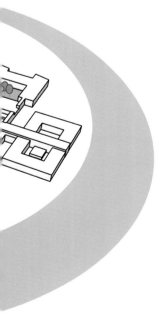

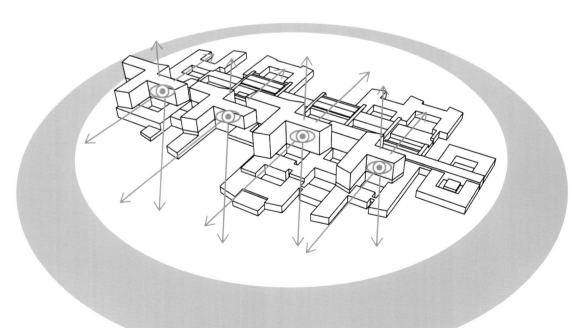

VIEWS FROM WARDS
病房的视野
Direct seaviews from all bed wards
从所有病房都能直接看到海景

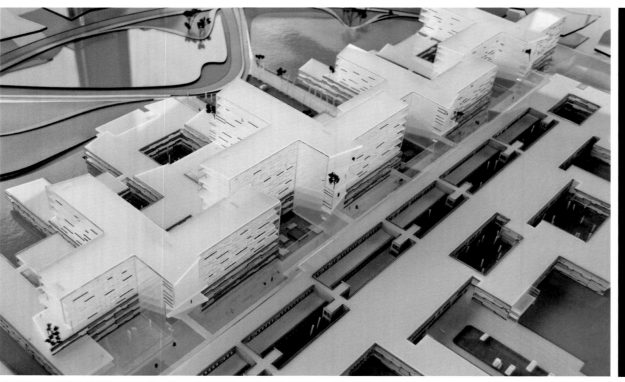

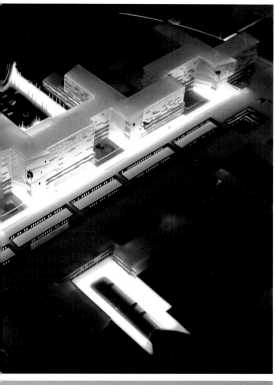

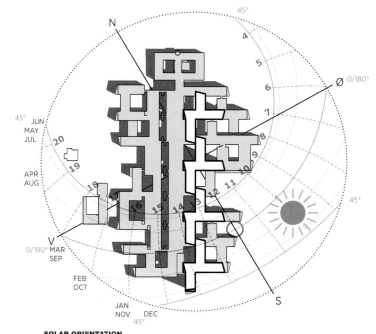

SOLAR ORIENTATION
太阳方位

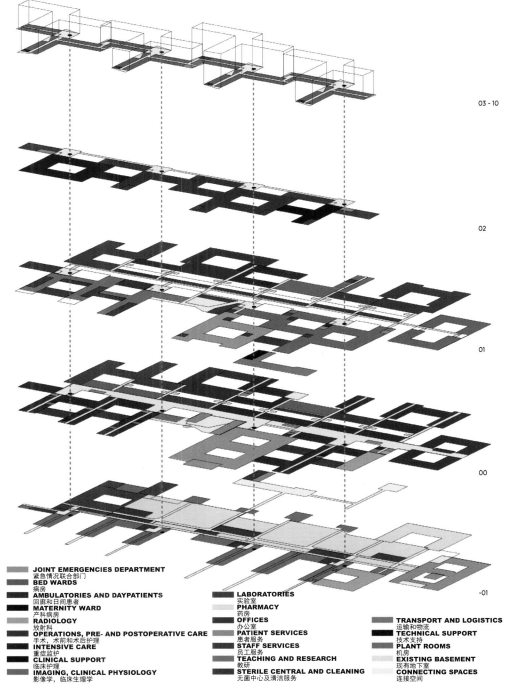

- **JOINT EMERGENCIES DEPARTMENT**
  紧急情况联合部门
- **BED WARDS**
  病房
- **AMBULATORIES AND DAYPATIENTS**
  回廊和日间患者
- **MATERNITY WARD**
  产科病房
- **RADIOLOGY**
  放射科
- **OPERATIONS, PRE- AND POSTOPERATIVE CARE**
  手术，术前和术后护理
- **INTENSIVE CARE**
  重症监护
- **CLINICAL SUPPORT**
  临床护理
- **IMAGING, CLINICAL PHYSIOLOGY**
  影像学，临床生理学
- **LABORATORIES**
  实验室
- **PHARMACY**
  药房
- **OFFICES**
  办公室
- **PATIENT SERVICES**
  患者服务
- **STAFF SERVICES**
  员工服务
- **TEACHING AND RESEARCH**
  教研
- **STERILE CENTRAL AND CLEANING**
  无菌中心及清洁服务
- **TRANSPORT AND LOGISTICS**
  运输和物流
- **TECHNICAL SUPPORT**
  技术支持
- **PLANT ROOMS**
  机房
- **EXISTING BASEMENT**
  现有地下室
- **CONNECTING SPACES**
  连接空间

墨西哥合众国墨西哥城"垂直公园"
Vertical Park

设计单位：Jorge Hernandez de La Garza
项目地址：墨西哥合众国墨西哥城
设计团队：Rodrigo Ambriz Michael Smith
　　　　　Erik Cosio

Designed by: Jorge Hernandez de La Garza
Location: Mexico City, Mexico
Design Team: Rodrigo Ambriz, Michael Smith, Erik Cosio

项目概况

项目位于墨西哥城科约阿坎区，这一现代化的项目不仅为人们提供了便利的生活和工作空间，同时也是一个生态的、可持续的城市空间。

设计理念

在墨西哥城这个大都市里，各区域交织成网，商业区与生活区融合，构建了一个相互作用、相互连接的动态空间。然而，随着现代化进程的推进，离岛式项目的出现使得这些动态空间退居城市边缘。各功能区的分离，既导致了交通的不便，同时也使越来越多的绿色空间被钢筋水泥所替代。在利益的驱使下，这种纯粹追求商业价值的开发模式对城市文化和城市的可持续发展构成了威胁。为此，该方案提出了"垂直花园"这一新的城市发展模式，以解决现代化进程中产生的问题。

设计特色

设计构建了一个高度结构化、极度灵活的体量，这一体量通过在水平方向上和垂直方向上的空间堆叠以及各式空间的嵌入，在城市外围形成一个相互关联的网格结构。体量的垂直上升形成了一个高度结构化的空间，其在水平方向上的延伸，则创造了一个有檐棚的商业街。同时，这一结构的高度灵活性使之能在整个墨西哥城乃至世界各地进行结构重组和改造。

Profile

Located in Coyoacan, Mexico City, the modernized project offers convenient living and working spaces as well as ecological sustainable urban spaces for the public.

Design Concept

Mexico City is a metropolis where zones are meshed as commerce and life often converge to create dynamic spaces of interaction and interconnection. Unfortunately, these zones are being pushed to the periphery as modern development increases delineation to create islands of program disconnected from the pulse of the city. With all its potential for profit, this march of capitalism poses a threat on the cultural and sustainable potential of the city. The proposal for "Vertical Park" is to revive the calcification of modernity.

Design Feature

In response to these demands, designers have designed a module, highly structural and flexible volume in order to provide horizontal and vertical stacking along with diverse insertions. As the modules rise vertically to create a high-rise structure, they also spread horizontally in order to create canopies for street level commerce. These modules can be rearranged, relocated, and remodeled throughout Mexico City and potentially throughout the world.

Vertical Park, Highly Flexible Volume

垂直花园 高度灵活性

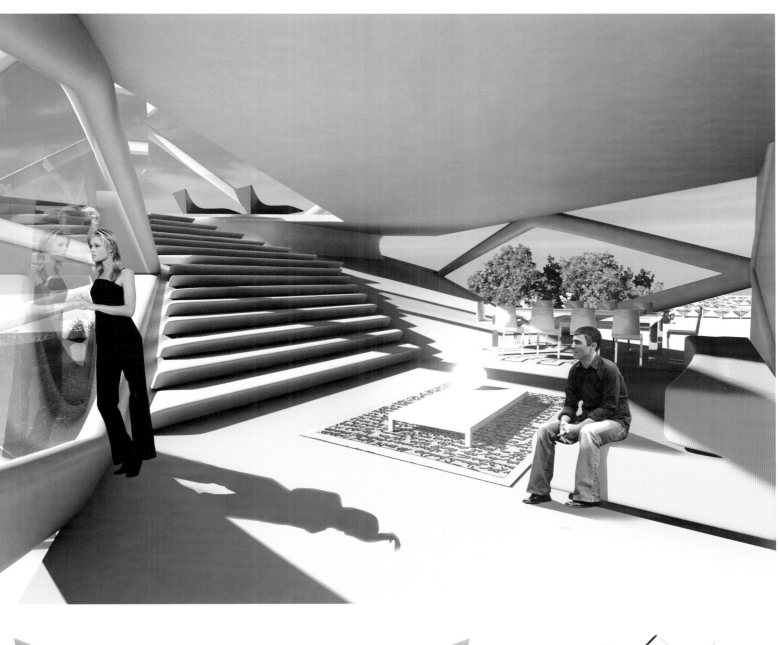

墨西哥合众国库利亚坎 Nativity 教堂
Nativity Church

设计单位：Pascal Arquitectos
项目地址：墨西哥合众国库利亚坎

Designed by: Pascal Arquitectos
Location: Culiacan, Mexico

"鱼"
野餐花园
通风立面

"Fish",
Picnic Garden,
Ventilated Façade

设计构思

在基督教中，"鱼"是一个古老而鲜明的标志符号，这不仅因为其希腊语"Ichtus"以离合诗的形式表示"耶稣基督"，也跟耶稣的教导跟鱼有关有着很大的关联。建筑的形态源自于鱼，建筑外表上呈一定序列排布的覆盖层也使人们很自然地联想到鱼鳞，成为极具基督教特色的标志。这一设想不仅赋予了建筑强烈的识别性，也使建筑蕴含了深刻的建筑内涵。

绿色建筑的诠释方式丰富多样，却建立在一个共同的基础上，即其设计和操作应尽可能降低对人体健康和自然环境的负面影响。本案中，设计致力于提升资源利用效率的同时，在建筑生命周期层面上，通过合理的选址、设计、施工、操作和维护，实现建筑、人、自然的共生。

设计特色

在功能布局上，考虑到造价和布局的合理性，设计师没有将野餐区设置在桥梁上，而是设置在草地上。整个野餐区围绕着广场外围分布，形成了一个野餐花园，与广场毗邻，却又自然地分隔开来，彼此间不会产生干扰。广场周围绿树成荫，形成了一道天然的绿墙，既创造了一个独立的场所，又与教堂的遮蔽构建共同营造出宜人的微气候环境。这一道绿墙也好似一层面纱，将周围的景观半遮掩起来，产生一种朦胧美。

设计充分考虑了库利亚坎的气候条件。建筑有着双层通风立面，外层表面可反射热量，其后的防潮层可使内外空气自由流动，形成了一个应对气象的保护屏障，既使入射阳光不会直接进入到建筑内部，又保证了建筑的自然通风。这一被动系统的采用，使建筑在夏季可以自然冷却，在冬季可以自然升温，既节约了机械通风设备和空调设备，也降低了建筑的能源消耗。

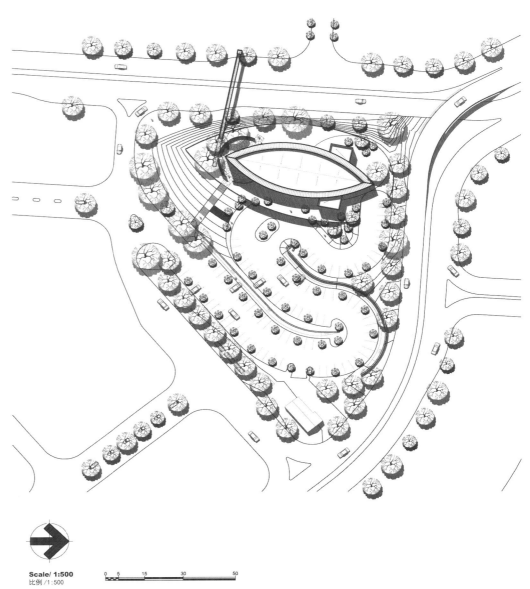

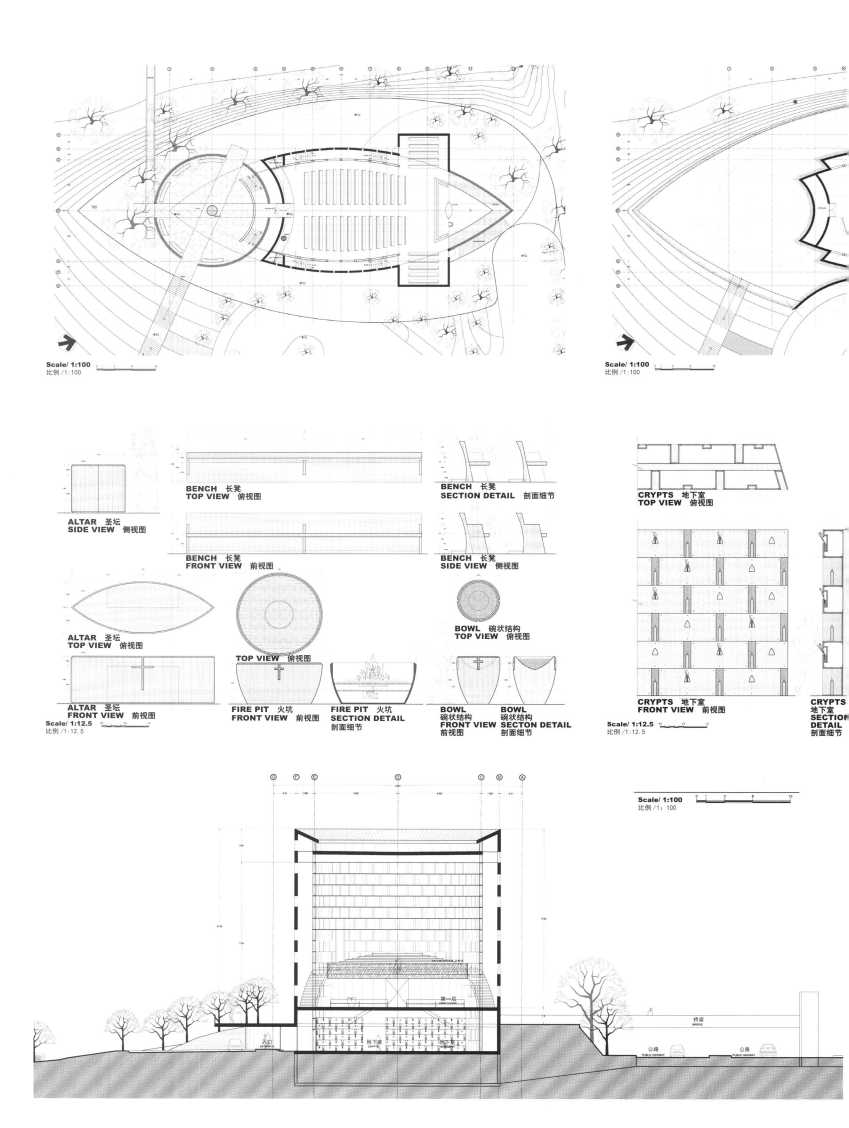

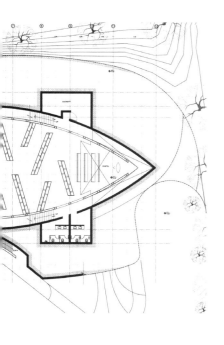
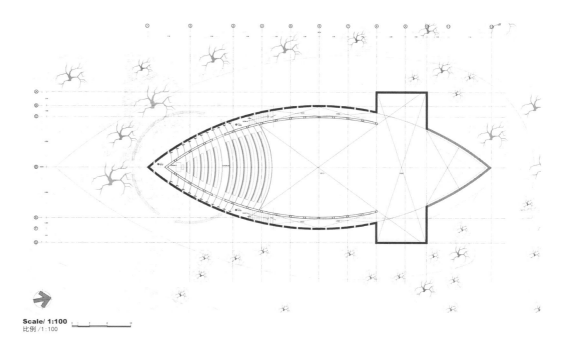

Scale/ 1:100
比例 /1:100

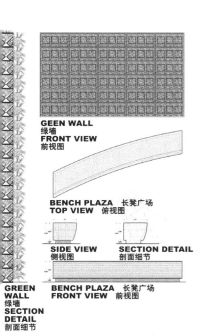

GEEN WALL
绿墙
FRONT VIEW
前视图

BENCH PLAZA 长凳广场
TOP VIEW 俯视图

SIDE VIEW | **SECTION DETAIL**
侧视图 | 剖面细节

GREEN WALL | **BENCH PLAZA** 长凳广场
绿墙 | **FRONT VIEW** 前视图
SECTION
DETAIL
剖面细节

EXTERIOR WALL DETAIL 外墙细节 | **EXTERIOR WALL** 外墙 **SECTION DETAIL** 剖面细节 | **STONE WALL DETAIL** 石墙细节 | **STONE WALL** 石墙 **SECTION DETAIL** 剖面细节

NATIVITY SCENE DETAIL
基督诞生场景细节
Scale/ 1:25
比例 /1:25

WALL SECTION DETAIL
墙体剖面细节

LATTICE INTERIOR DETAIL
格架内部细节

LATTICE
格架
SECTION DETAIL
剖面细节

Scale/ 1:100
比例 /1:100

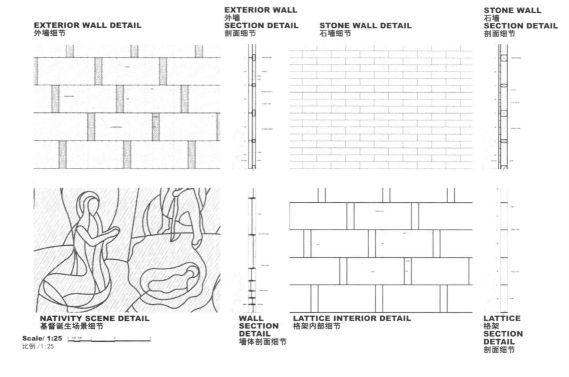
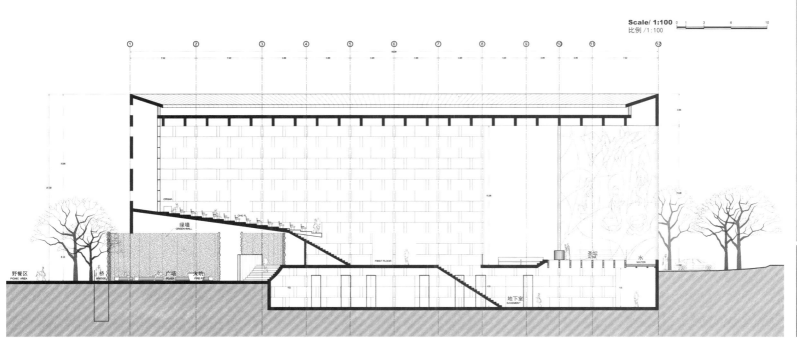

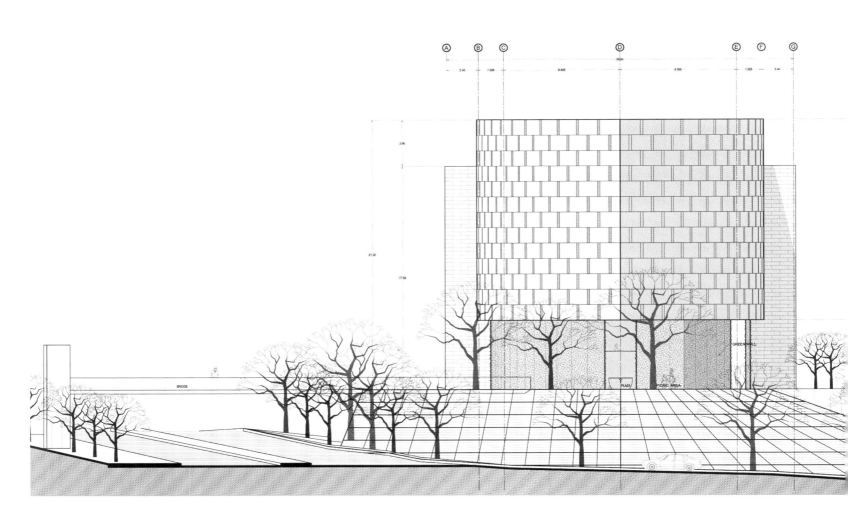
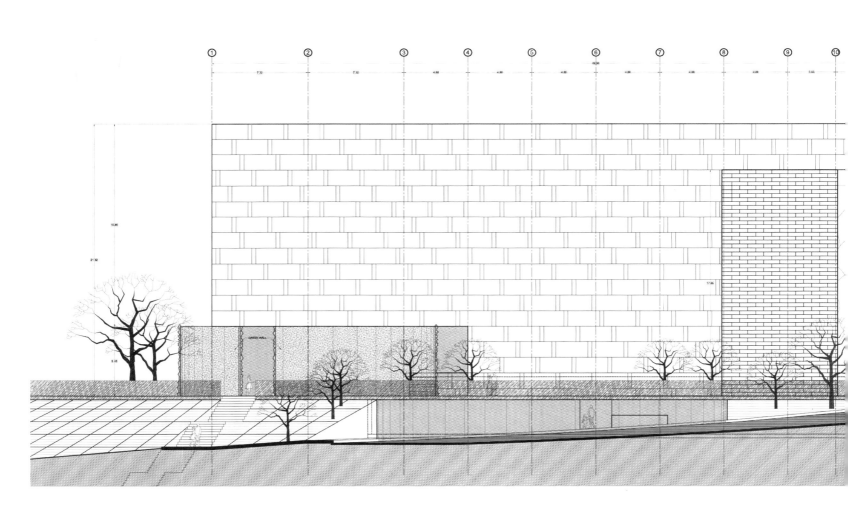

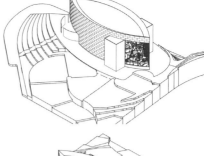
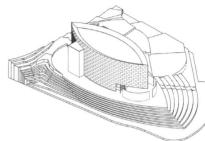

Scale/ Without Scale
比例 / 不成比例

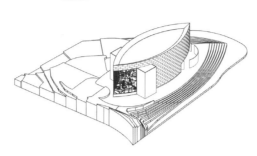
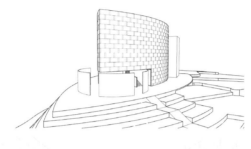
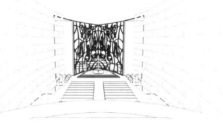

Scale/ Without Scale
比例 / 不成比例

Design Concept

In the Christian, "Fish" is an ancient distinct symbol not only because "Ichtus" represents "Jesus Christ" in acrostic but also because Christ's teachings are closely connected with fish. The project's intention is to create a recognizable object that its shape comes from a fish, and the exterior covering reminds of scales. It will give the building strong identity and profound architectural connotation.

Though green building is interpreted in many different ways, a common view is that they should be designed and operated to reduce the overall impact of the built environment on human health and the natural environment. The design philosophy focuses on increasing the efficiency of resource use while reducing building impacts on human health and the environment during the building's lifecycle, through better siting, design, construction, operation, maintenance, and removal, achieving the coexistence of architecture, man and nature.

Design Feature

For the consideration of cost and a rational layout, designers decided not to put the picnic area in the bridge but on the grass. A picnic garden outside the plaza has a view to the lake and it is shaded by big trees. It adjoins the plaza but not together with the plaza events. The green wall in the plaza creates a contained place. Together with the shadow of the church it generates a microclimate. The green wall acts like a veil partially covering surrounding landscape.

The design has fully considered the Culiacan weather condition. The building is going to be surrounded by a double ventilated façade, in which the exterior cladding reflects the heat, behind there is a vapor barrier which allows the passage of air, so that the incidence of the sun is not direct into the building, and it also creates a meteorological barrier. With this passive system, the building is naturally cooled in summer and naturally heated in winter. It decreases the use of mechanical ventilation and conditioning equipment whilst reduces the energy consumption of the building.

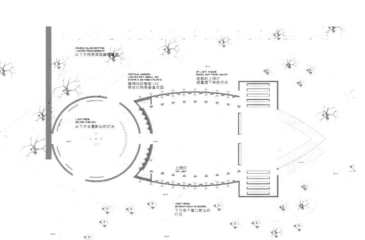

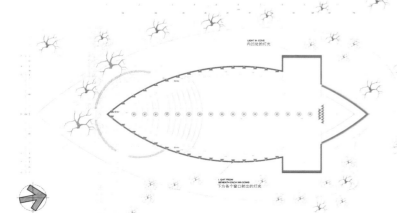

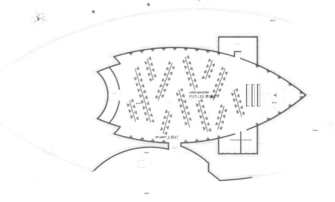

SYMBOLOGY
符号

▽ **WALL SCONCE** 壁灯
▽ **STEP LIGHT** 台阶灯
▼ **DOWN LIGHT** 下照灯
▼ **UP LIGHT** 上照灯

Scale/ 1:200
比例 /1:200

公司简介 Company Profile

阿特金斯 ATKINS

阿特金斯是世界领先的设计顾问公司之一，专业知识的广度与深度使之能够应对具有技术挑战性和时间紧迫性的项目。阿特金斯是国际化的多专业工程和建设顾问公司，能为各类开发建设项目提供一流的专业服务，从摩天大楼设计到城市规划、铁路网络改造，以及防洪模型的编制，都能提供规划、设计、实施的全程解决方案。

作为阿特金斯集团远东区的全资子公司，阿特金斯中国于 1994 年正式进入中国市场。借助集团总部的强大支持，阿特金斯中国以提供多专业多学科的"一站式"全方位服务的核心优势区别于竞争对手，并通过国际经验和本地知识的有机结合，在中国近年来迅猛推进的城市化进程中取得了骄人的项目业绩。

其主要作品和奖项有：中央景城一期荣获 2010 年中国土木工程詹天佑奖住宅小区优秀规划奖；伊顿小镇荣获 2010 年中国土木工程詹天佑奖住宅小区金奖；上海办事处荣获 LEED 绿色建筑商业内部装修金奖等多个奖项。

Atkins

Atkins is one of the world's leading design consultancies. It has the breadth and depth of expertise to respond to the most technically challenging and time-critical projects and to facilitate the urgent transition to a low carbon economy. Atkins' vision is to be the world's best design consultant. Whether it's the architectural concept for a new supertall tower, the upgrade of a rail network, master planning a new city or the improvement of a management process, designers plan, design and enable solutions.

With 75 years of history, 17,700 employees and over 200 offices worldwide, Atkins is the world's 13th largest global design firm (ENR 2011), the largest global architecture firm, the largest multidisciplinary consultancy in Europe and UK's largest engineering consultancy for the last 14 years. Atkins is listed on the London Stock Exchange and is a constituent of the FTSE 250 Index.

In 1994 Atkins established its first Asian office in Hong Kong followed by Singapore in 1996. Today Atkins also has offices in Hong Kong, Beijing, Shanghai, Chengdu, Chongqing and Ho Chi Minh and Sydney, all part of an integrated network that delivers innovative multidisciplinary projects and employs approximately 1,000 staff across the region from China to South East Asia and Australia.

Architekt Lukas Göbl | Office for Explicit Architecture

Architekt Lukas Göbl | Office for Explicit Architecture 致力于为各类型的建筑设计和城市规划项目提供从草图设计到设计构思原理、场址及材料的选择、现场施工监督等全方位的服务。其设计的独特性和个性源自对现实与理想的认识、大胆的构思、自由的思想以及精细的细部处理。在设计中，他们将传统绘画技术与现代传媒系统结合起来，综合考虑各方面的因素，使方案最优化。

Architekt Lukas Göbl | Office for Explicit Architecture

Architekt Lukas Göbl | Office for Explicit Architecture works on all types of architecture and urban planning, providing comprehensive services in these areas. The spectrum ranges from the first decisive sketch and the formulation of the design philosophy to the careful choice of location and materials, meticulous planning, and project specification all the way to on-site construction supervision. The uniqueness and individuality of its works arises from the coexistence of vision and reality, boldness and precision, free spirit and exact detail. Traditional drafting techniques and new media are coordinated in the design, as well as a comprehensive consideration of all possible perspectives, to achieve maximum optimization.

Dominique Perrault Architecture

Dominique Perrault Architecte 于 1981 年由 Dominique Perrault 在巴黎建立。事务所通过不断的研究和创新性设计推动建筑设计、城市规划和设计的发展。该事务所重新定义了"建筑"这一词汇，其对城市建筑新类型以及建筑材料（包括金属material 的使用）的研究，使之在建筑研究上起着领军作用。

公司创始人 Dominique Perrault 1953 年出生在法国的克拉蒙特，曾获得过法国国立高等美术学院的建筑学学位、巴黎国立路桥学院城市规划高级文凭、巴黎社会科学高等学院历史学硕士文凭。同时，他也是法国荣誉爵士、法国建筑学会的成员、德国建筑师协会的荣誉会员（BDA）和英国皇家建筑学会的成员。Dominique Perrault 获得过多项大奖，包括 1996 年法国国家建筑奖、1997 年因国家图书馆的设计而获得的密斯·凡·德·罗奖。

Dominique Perrault Architecture

DPA is conceived as a laboratory of research and innovation to push the natural boundaries of architecture, urban planning and design. DPA LAB develops processes reinventing the vocabulary of architecture, leading a major research role not only on materials – including the early use of the metal mesh – but in new typologies of urban approaches.

Dominique Perrault (1953, Clermont-Ferrand) is a French architect. He became world known for the design of the French National Library, distinguished with the Mies van der Rohe Prize in 1996.

He received his Diploma in Architecture at the Ecole Nationale Supérieure des Beaux-Arts in Paris in 1978. He also holds postgraduate diplomas in Town Planning from the Ecole supérieure des Ponts et Chaussée and History from the Ecole des Hautes Etudes en Sciences Sociales. He currently heads Dominique Perrault Architecture (DPA) in Paris.

Cervera & Pioz Arquitectos

Cervera & Pioz Arquitectos 成立于 1979 年，由设计师 Javier Pioz 和 Rosa Cervera 共同创立。该公司致力于构建独特而又具有标志性意义的建筑，其设计作品获得多项荣誉和奖项。

自公司成立后，Cervera 和 Pioz 在对逻辑原则的灵活性与适应性以及能源的高效利用等方面有了全面理解的基础上，为提出新的建筑设计和城市规划方案而奋斗。

1992 年，Cervera 和 Pioz 开始了将生物工程应用到现代结构工程和建筑设计中的研究，并由此提出一种工程结构和设计的新概念——生物结构。继而，他们将这一观念运用到高层空间中，发展可持续的垂直生物结构工程，这是一种可持续的、以人为本的城市发展新模型。这一理论强调对自然规律的尊重和利用，以实现自然与技术之间的平衡，在国际上具有重大的影响力。

Cervera & Pioz

Founded by Javier Pioz and Rosa Cervera in 1979, the Firm has, since then, devoted itself to the creation of unique and emblematic buildings. Their work has won many distinguished awards and prizes, including:

Administrative Centre of the Spanish town in Shanghai (1); High School Complex (1); Xixi Wetland Museum in Hangzhou (2); P.R. of China Embassy in Madrid (1); Citibank-Spain Competition in Cadiz (1); Islamic Cultural Centre in Madrid (3); the National Festival of Video-Art in Cadiz (2); Official College of Architects of Madrid Award; Antonio Maura Award from the City Council of Madrid; Diploma from the Architecture Biennial in Sofia, Bulgaria; and Honor Award by the "Foundation For Architectural & Environmental Awareness" and "ArchiDesign Perspective", New Delhi, India.

LAVA

LAVA 是由 Chris Bosse、Tobias Wallisser 和 Alexander Rieck 于 2007 年创立的一家跨国公司，它是创造性研究和设计思维的交汇，在悉尼、上海、斯图加特和阿布扎比设有办事处。其业务包括了城市中心总体规划、家具、酒店、住宅以及机场设计。

LAVA 热衷于从雪花、蜘蛛网、肥皂泡沫等自然形态中探索新的建筑类型和结构，这种源自自然的几何体实现了功能与美观的统一。LAVA 擅长将智能系统和建筑外皮融合，以此对空气压力、温度、湿气、太阳辐射、污染等外界因素作出反应。

LAVA 获得的主要奖项有：澳大利亚室内设计奖、多伦多"零脚印公司"颁发的国际建筑表面改造奖、《国际设计》杂志年度设计回顾奖和 IDEA 奖、英国伦敦建筑设计大奖、迪拜都市景观建筑奖。该公司还曾获得 Iakov Chernikhov 国际大奖、2011 年 Dedalo Minosse 国际建筑奖和科技创新奖提名。

LAVA

Chris Bosse, Tobias Wallisser and Alexander Rieck founded multinational firm, Laboratory for Visionary Architecture (LAVA) in 2007 as a network of creative minds with a research and design focus and with offices in Sydney, Shanghai, Stuttgart and Abu Dhabi. The business scope of LAVA covers urban center master plan, furniture, hotel, houses, airports, etc.

The potential for naturally evolving systems such as snowflakes, spider webs and soap bubbles for new building typologies and structures has continued to fascinate LAVA – the geometries in nature create both efficiency and beauty. LAVA's projects incorporate intelligent systems and skins that can react to external influences such as air pressure, temperature, humidity, solar radiation and pollution.

Awards include the Australian Interior Design Awards, UN partnered ZEROprize Re-Skinning Award, I. D. Annual Design Review, IDEA Awards, AAFAB AA London, Cityscape Dubai Award Sustainability; commendations include Well Tech Award and Dedalo Minosse International Prize in 2011; and nominations for the Iakov Chernikhov International Prize and the Index Award.

HKG Group

HKG Group 创立于 1952 年，以温哥华为发展基地，其业务已遍及全球，在过去数十年间，一直是北美建筑设计业最具规模的前 100 家设计公司之一。自 2000 年起，为进一步拓展亚太市场，HKG 以中国上海为其另一设计基地，并于 2001 年成立了 HKG 上海办事处，融入了东西方的文化，吸纳了一批极富理想和创意的优秀设计师。

HKG 曾经为世界各地的顾客设计过机场交通、大型办公、高端酒店、商业综合广场及宗教文化等项目。在众多共同的价值观基础上，HKG 顺应客户规划、需求及特质，充分发挥创造性，提出了众多充满创意的设计方案。

HKG Group

HKG Group established in 1952 has Vancouver as its development base and expands its business worldwide. For over 20 years, HKG has been included in the industry rankings of the top 100 and 200 design firms in North America. Since 2000, to expand Asia-Pacific Market, HKG decided to set another base in Shanghai and then established the HKG Shanghai Office. As a joint venture company in Shanghai, HKG Shanghai Office has created a fusion working environment where Eastern & Western culture & principles are merged harmoniously.

The business of HKG includes airport & transportation projects, commercial office building, hospitality, retail & entertainment, and cultural centers.

SAMYN and PARTNERS, Architects & Engineers

SAMYN and PARTNERS, Architects & Engineers 是一家由 Ir Philippe SAMYN 博士领导的私人公司。随着其附属公司 Ingenieursbureau Jan MEIJER、FTI、DAE、AirSR 的相继建立，该公司也在建筑设计和建筑工程各领域表现得极为活跃。其业务范围涵盖了规划设计、城市规划、景观设计、建筑设计、室内设计、建筑物理、MEP 和工程结构、工程建设管理、成本规划与控制、工程造价管理等多方面。

SAMYN and PARTNERS, Architects & Engineers 的设计方案建立在"质疑"的基础上，可用"为什么"理念来概括。该公司尝试着接手各种类型的项目，并悉心听取客户的意见和需求。

SAMYN and PARTNERS, Architects & Engineers

SAMYN and PARTNERS, Architects & Engineers is a private company owned by its partners and lead by its design partner Dr Ir Philippe Samyn. With the establishment of its affiliated companies Ingenieursbureau Jan MEIJER, FTI, DAE, and AirSR, it is active in all fields of architecture and building engineering. The firm's client services include Planning and Programming, Urban Planning, Landscaping and Architectural Design, Interior Design, Building Physics, MEP and Structural Engineering, Project and Construction Management, Cost and Planning Control, Quantity Surveying, Safety and Health Coordination.

Philippe Samyn's architectural and engineering design approach is based on questioning, which can be summarized as a "why" methodology. The firm approaches projects openly to all sorts of possibilities whilst listening closely to its clients' demands.

Henning Larsen 建筑事务所

Henning Larsen 建筑事务所于 1959 年由建筑师 Henning Larsen 创立，是一家根植于斯堪的纳维亚文化的国际建筑公司。Henning Larsen 建筑事务所由首席执行官 Mette Kynne Frandsen 以及设计总监 Louis Becker 和 Peer Teglgaard Jeppesen 负责管理，已在哥本哈根、慕尼黑、贝鲁特以及利雅得建立了办事处。

Henning Larsen 建筑事务所在环境友好型建筑和综合节能设计等方面卓有建树，以创造宜人、可持续的项目为目标，给当地的人们、社会及文化创造持久的价值。设计师对项目负有高度的社会责任感，注重与客户、业主以及合作者的交流，积累了从建议书草案到细节设计、监督和施工管理等各方面的建筑知识和理论。

其主要作品有：在都柏林设计的"钻石"大厦、哥本哈根商业学校、瓦埃勒"波浪住宅"、格鲁吉亚巴统水族馆、瑞典于默奥（Umea）大学建筑学院楼等。

Henning Larsen Architects

Henning Larsen Architects is an international architecture firm based in Copenhagen, Denmark. Founded in 1959 by noted Danish architect and namesake Henning Larsen, it has around 200 employees. In 2011 the company worked on projects in more than 20 countries.

In 2008 Henning Larsen Architects opened an office in Riyadh, Saudi Arabia named Henning Larsen Middle East and in 2011 an office in Munich, Germany was inaugurated. Most recently Henning Larsen Architects opened two offices, one in Oslo, Norway and one in Istanbul, Turkey.

Henning Larsen Architects is known for their cultural and educational projects. Last year Harpa Concert Hall and Conference Centre in Reykjavik was selected as one of the ten best concert halls in the world by the British magazine Gramophone. Henning Larsen Architects also designed the Copenhagen Opera.

Current projects include a new headquarter for Siemens in Munich, Germany and a 1.6 mill m² masterplan for the King Abdullah Financial District in Riyadh, Saudi Arabia.

斯蒂文·霍尔建筑师事务所

斯蒂文·霍尔建筑师事务所是一所由斯蒂文·霍尔（Steven Holl）于 1976 年创立的建筑设计事务所，业务范围包括建筑设计与城市规划。事务所擅长有关艺术和高等教育类型的建筑设计，作品包括赫尔辛基当代美术馆、纽约普拉特学院设计学院楼、爱荷华大学艺术与艺术史学院楼、西雅图圣伊格内修斯小教堂。

斯蒂文·霍尔建筑师事务所在国际上享有盛誉，以其高质量的设计多次获奖，其作品多次获得出版或展览，所获奖项有：北京当代 MOMA 荣获由国际高层建筑与城市住宅协会（CTBUH）所颁发的"2009 年世界最佳高层建筑"大奖、2008 年美国建筑师协会纽约分会可持续设计大奖、2009 年西班牙对外银行（BBVA）基金会知识前沿奖、尼尔森·阿特金斯美术馆获得 2008 年美国建筑师协会纽约分会建筑荣誉大奖。斯蒂文·霍尔本人被授予"美国建筑师协会金奖"，这个奖项旨在认可对建筑界做出持久贡献的个人。

Steven Holl Architects

Steven Holl Architects is a 40-person innovative architecture and urban design office working globally as one office from two locations; New York City and Beijing. Steven Holl leads the office with senior partner Chris McVoy and junior partner Noah Yaffe. Steven Holl Architects is internationally-honored with architecture's most prestigious awards, publications and exhibitions for excellence in design. Steven Holl Architects has realized architectural works nationally and overseas, with extensive experience in the arts (including museum, gallery, and exhibition design), campus and educational facilities, and residential work. Other projects include retail design, office design, public utilities, and master planning.

Steven Holl Architects has been recognized with architecture's most prestigious awards and prizes. Most recently, Steven Holl Architects' Cite de l'Ocean et du Surf received a 2011 Emirates Glass LEAF Award, and the Horizontal Skyscraper won a 2011 AIA National Honor Award. The Knut Hamsun Center received a 2010 AIA NY Honor Award, and the Herning Museum of Contemporary Art received a 2010 RIBA International award. Linked Hybrid was named Best Tall Building Overall 2009 by the CTBUH, and received the AIA NY 2008 Honor Award. Steven Holl Architects was also awarded the AIA 2008 Institute Honor Award and a Leaf New Built Award 2007 for The Nelson-Atkins Museum of Art (Kansas City).

Process-based Architecture Studio

Process-based Architecture Studio，简称为 PA Studio，由 Jafar Bazzaz, Arash Pouresmaeil 和 Kamal Youssefpoor 建立于 2010 年，此三人均毕业于 IAUT 的建筑专业。PA Studio 探索了不同的空间理念和空间转化策略，以为各类型的建筑设计提供别出心裁的解决方案。在每一个项目中，他们都会确定一个建筑要点来定义他们在不同案例中采用的空间理念和转化策略。"规划、环境、空间形态"是该工作室设计理念的主要来源，重视加工则是该工作室的基本原则，因为在他们看来，设计加工与设计是同等重要的。

Process-based Architecture Studio

PA Studio | Process-based Architecture Studio was founded in 2010 by Jafar Bazzaz, Arash Pouresmaeil and Kamal Youssefpoor. They graduate in architecture from Islamic Azad University of Tabriz. PA Studio explores different spatial concepts and transformation strategies to find inventive and creative solutions for architectural problems. Program, Context and Spatial Patterns are main sources of architectural ideas in different PA Studio projects. All projects have an architectural keyword that defines their spatial concept and transformation strategy. Emphasis on design process is the most fundamental principle in PA Studio projects. PA Studio believes that design process is as important as design.

C.F.Møller Architects

C.F.Møller Architects 是斯堪的纳维亚半岛历史最悠久、规模最大的建筑机构之一，业务范围涵盖方案分析、城市规划、总体规划、景观设计、建筑工程设计及其他诸多领域。事务所于 1924 年创立，以简约、明快、朴素的理念指导各项实践，并根据每个项目的基地特点，结合国际发展趋势和地域差异对理念进行重新解读和诠释。

事务所以改革和创新为发展理念，力图打造独具吸引力和发展前景的工作环境，使每位员工都能接受高要求设计项目的挑战。多年来，事务所屡获国内外设计大奖，其作品多次在国内外的展会上展出。其主要作品有：法尔斯特岛新封闭式州立监狱、奥尔胡斯 Incuba 科学园、奥尔胡斯大学礼堂、奥尔胡斯艺术大楼扩展项目、奥尔胡斯低能耗办公大楼、国家海事博物馆扩建、巴里考古博物馆等。

C.F. Møller Architects

C.F. Møller is one of Scandinavia's oldest and largest architectural practices. Its award-winning work involves a wide range of expertise that covers all architectural services, landscape architecture, product design, healthcare planning and management advice on user consultation, change management, space planning, logistics, client consultancy and organisational development.

Simplicity, clarity and unpretentiousness, the ideals that have guided its work since the practice was established in 1924, are continually re-interpreted to suit individual projects, always site-specific and based on international trends and regional characteristics.

C.F. Møller regards environmental concerns, resource-consciousness, healthy project finances, social responsibility and good craftsmanship as essential elements in its work, and this holistic view is fundamental to all its projects, all the way from master plans to the design.

Today C.F. Møller has about 320 employees. The head office is in Aarhus, Denmark and it has branches in Copenhagen, Aalborg, Oslo, Stockholm and London.

10 DESIGN（拾稼设计）

10 DESIGN（拾稼设计）是一家国际性的建筑设计事务所，其设计中心分别设于中国香港、中国上海、英国爱丁堡及美国丹佛。事务所的业务范围包括总体规划、建筑设计、建筑可持续性的研究、景观园林设计及室内设计。其作品涵盖大规模的城市规划、综合性功能开发、公共文化建筑、度假及酒店设施、企业总部、科研及办公设计及高端住宅设计等类型。

艾高登（Gordon Affleck）是拾稼设计的创始人之一，拥有超过 18 年在亚洲、英国、美国和中东等地带领国际设计团队的工作经验。他深信绿色低碳的可持续性技术与建筑的强力结合将成为建筑设计及与客户建立紧密合作关系的核心原则，这也是确保每个项目成功的关键。艾高登负责设计的已建成的地标性作品有：北京 2008 奥运会多用途场馆、关联场馆设计（现为中国国家会议中心）、南京华为研发总部园区等。

10 DESIGN

10 DESIGN is a leading International partnership of Architects, Urbanists, Landscape Designers and Animators. It works at all scales and sectors, including corporate, cultural, hospitality, retail, education and residential. 10 DESIGN's Architecture evolves from the multicultural nature of clients, and is driven by their respective social, economic and ecological conditions. 10 DESIGN has offices in Hong Kong, Edinburgh, Shanghai, Istanbul and Dubai. The current projects include the Central Business District in Chongqing, China; the Fujian Photonic University in Xiamen, China; Badshahpur Business Park in Gurgaon, India; Mavişehir Residential Development in Izmir; and the Route De Jussy 22 Thônex Project in Geneva, Switzerland. Projects under construction include Chengtou New Jiangwan Business Park in Shanghai, China; Bohai Bank Headquarters in Tianjin, China; and the Fujian Photonic University in Xiamen, China.

Oppenheim Architecture+Design

Oppenheim Architecture+Design（OAD）是一家提供建筑设计、室内设计和城市规划等全方位服务的公司。其总部位于美国佛罗里达州迈阿密市，在洛杉矶、瑞士巴塞尔设有办事处。公司由 Chad Oppenheim 创立，专注于为富有挑战性的复杂项目提供强有力的实用方案，拥有设计世界级医院、住宅和多用途建筑的丰富经验。

该公司从环境和相关规划中汲取精华，以创造一种戏剧性的、强有力的体验，同时也赋予建筑舒适感。其设计首先着眼于对客户需求的全方位分析，从而构建人性化的建筑和环境。该公司的主要作品有：Dellis Cay 别墅群、拉斯维加斯的 Hard Rock、索尼斯达毕士倾岛酒店、马可岛万豪酒店、哥伦比亚特区和亚特兰大酒店等。

Oppenheim Architecture + Design

Oppenheim Architecture + Design (OAD) is a full service architecture, interior design and urban planning firm located in Miami, Florida with offices in Los Angeles and Basel, Switzerland. The firm, founded by Chad Oppenheim, specializes in creating powerful and pragmatic solutions to complex project challenges and has extensive experience in world class hospitality, residential and mixed-use design.

The firm's design strategy is to extract the essence from each context and relative program – creating an experience that is dramatic and powerful, yet simultaneously sensual and comfortable. The firm's approach begins by a comprehensive analysis of a client's vision in relation to the projects typology, context, zoning parameters and financial realities.

Since its inception, OAD has accumulated over $100 billion in project work and has won over 45 awards and honors for its unique design sensibility and ability to pioneer innovative concepts, optimize challenging sites and revitalize blighted urban areas. Projects designed by the firm cover a broad spectrum of programmatic requirements, budgets and building types. This award-winning work is based on both a physical and spiritual contextual sensitivity, supported by evocative and economic design solutions. More specifically, the firm's pragmatic yet poetic architectural solutions in unproven and undeveloped areas have served to revitalize many neighborhoods throughout the world.

TOTEMENT/PAPER

TOTEMENT/PAPER 成立于 2006 年，是一家年轻的建筑公司。该事务所将其工作视为一种研究活动，建筑对他们来说，不是一种工具、一种目的，而是作为与其他人类活动方式的一个扩展过程。事务所结合文化、哲学、伦理、技术等多学科知识，试图表现属于自己的空间美学。TOTEMENT/PAPER 自成立以来参加了多项国内外的设计比赛，曾获得多个奖项。其设计团队还参与了 2010 年上海世博会俄罗斯展馆的设计。

TOTEMENT/PAPER

TOTEMENT / PAPER – the young architectural command organized in 2006 which considers as a basis of the activity "research", but not as the tool of reception of result, and as process of expansion of communications with other kinds of human activity where the architecture is not the purpose and result, and the tool, way and an esthetics of language of dialogue. On a joint of various languages (cultural, technological, philosophical, ethical etc.), designers try to find their own esthetics of "space-form" which can become a basis for future ethics of communications.

Bureau TOTEMENT / PAPER during the existence repeatedly took part in the Russian and foreign competitions, winning the first places, becoming participants of short lists and winners. Bureau works are published in the Russian and international editions, takes active part in exhibition and lecture activity. TOTEMENT/PAPER is the author of Russian pavilion at EXPO-2010 in Shanghai.

de Architekten Cie.

de Architekten Cie. 是一家全球性的建筑设计公司，具有 30 多年的建设设计和规划经验。其主要业务范围包括总体规划、城市规划、建筑设计和室内设计。

公司的主要创始人 Pi de Bruijn 1967 年毕业于代尔夫特理工大学建筑学院，其后分别在伦敦萨瑟克区的伦敦市委员会建筑系和阿姆斯特丹市政房屋署工作，并于 1978 年成为 Oyevaar Van Gool De Bruijn 建筑事务所 BNA 办事处的合伙人。1988 年，Pi de Bruijn 和 Frits van Dongen, Carel Weeber, Jan Dirk Peereboom Voller 共同建立了 de Architekten Cie.。

de Architekten Cie.

Branimir Medić & Pero Puljiz, de Architekten Cie. is a global architectural design company with over 30 years of architectural design and planning experience. The business scope covers overall planning, urban planning, architectural design and interior design. Pi de Bruijn, the chief founder of the company, completed his studies at the Faculty of Architecture at Delft University of Technology in 1967. He then left to work at the Architects Department of the London City Council in Southwark, London. On his return to Amsterdam, he worked for the Municipal Housing Department, until he established himself as an independent architect in 1978, as a partner in the Oyevaar Van Gool De Bruijn Architecten BNA bureau. In 1988 he founded de Architekten Cie. together with Frits van Dongen, Carel Weeber and Jan Dirk Peereboom Voller, and has been a partner ever since.

Architetto Michele De Lucchi S.r.l

Michele De Lucchi 1951 年出生于意大利菲拉拉，后毕业于意大利佛罗伦萨的建筑学院。1992 年 – 2002 年，Michele De Lucchi 担任意大利百年品牌公司 Olivetti 的设计总监，在此期间，他获得与惠普、飞利浦、西门子等电子巨头的合作机会。

设计师的创意与专业水平来自于其对建筑、设计、科技、以及工艺的自我探索。其主要作品有：米兰三年展中心、罗马展览博物馆、柏林新博物馆、意大利的 le Gallerie、米兰斯卡拉广场、Radison 酒店、巴统服务大楼、第比利斯和平桥等。

Architetto Michele De Lucchi S.r.l

Michele De Lucchi was born in 1951 in Ferrara and graduated in architecture in Florence. During the period of radical and experimental architecture he was a prominent figure in movements like Cavart, Alchymia and Memphis. De Lucchi has designed furniture for the most known Italian and European companies. From 1992 to 2002, he has been Director of Design for Olivetti and developed experimental projects for Compaq Computers, Philips, Siemens and Vitra.

His professional work has always gone side-by-side with a personal exploration of architecture, design, technology and crafts. He designed buildings for museums including the Triennale di Milano, Palazzo delle Esposizioni di Roma, Neues Museum Berlin and the le Gallerie d'Italia Piazza Scala in Milan. In the last years he developed many architectural projects for private and public client in Georgia, that include the Ministry of Internal Affairs and the bridge of Peace in Tbilisi, the Radison Hotel and Public Service Building in Batumi.

绿舍都会

绿舍都会（SUREArchitecture）2006 年成立于英国伦敦，并于 2008 年在北京设立分部，是一家专注可持续城市更新与生态建筑的国际化设计公司。它既是一家建筑设计公司，同时也是一个学术研究机构，与清华大学、谢菲尔德大学和其他专业机构的专家建立密切合作关系，尝试通过学术研究来激发建筑实践和创作。绿舍都会拥有一支来自不同背景的专业设计和研究团队，这使国际教育背景和国际工作经验成为该公司最大的优势和立足点。

SURE Architecture

SURE Architecture (Sustainable Urban Regeneration and Eco-Architecture) was established in London, UK in 2006 and had a branch in Beijing in 2008. It is a design and academic research architecture practice. The office would develop the research methodology for architecture theory and experiment to concentrate on sustainable urban regeneration and eco architecture design. The studio established close relations with many experts in Tsinghua University, University of Sheffield and other professional institutes in both China and UK. It would try to use academic research to motivate architecture experiment and creation. SURE Architecture has a team of professionals who are dedicated to architectural design and research. Their diverse international education background and work experience are distinctive advantages to provide top-quality services.

Brooks+Scarpa Architects

Brooks+Scarpa Architects 始建于 1991 年，其前身是 Pugh + Scarpa。对于建筑设计，该事务所认为该让用户尽可能地融入到建筑当中，增强其自主意识以及对建筑的认识与了解，给他们留下深刻的印象。事务所独立创造和探索，在他们看来，每一个作品都是研究的延续，也正是这种工作态度和方法，推动他们创作了众多创意作品，为每一个客户构建了专属的建筑。

Brooks + Scarpa 致力于保护环境，他们认为，全球社会正在经历一个态度的转变，即从"在地球上生存下来"转变为"改变我们的生存方式"。为此，他们积极探索对自然材料和文化资源的合理利用，通过建筑来改变人们的生活方式，从而推动社会的可持续发展。

Brooks + Scarpa Architects

Brooks + Scarpa is the successor architecture firm formerly known as Pugh + Scarpa which was established in 1991. Brooks + Scarpa believes that architecture should engage the users, heighten their sense of awareness, and bring a deeper understanding and vitality to their experience. The company strives to create environments that stimulate their occupants and leave lasting impressions. Brooks + Scarpa approaches each project as the continuation of an ongoing inquiry. This belief has produced a constant stream of inventive work leaving each client with a building which is solely their own. Brooks + Scarpa is committed to conserving the environment and making intelligent use of natural and cultural resources. They believe that societies across the globe are experiencing a shift in attitude from one merely concerned with surviving on earth to one concerned with changing how people live so that the earth can survive.

Casanova + Hernandez Architects

2001 年，Casanova + Hernandez Architects 在鹿特丹市成立，其业务范围只涉及城市规划、景观建筑和建筑设计三个相互关联的领域，旨在通过构建可持续发展的城市项目，为 21 世纪的人们营造新的城市居住环境。该事务所承接的项目，无论是一个地区或一个城市的大规模项目，还是与建筑空间或技术细节相关的小项目，其设计方案都是建立在对项目全面了解的基础上。

Casanova + Hernandez Architects

In 2001, Casanova + Hernandez Architects was established in Rotterdam. The office activities are focused on urban planning, landscape architecture, and architecture – the three linked fields to make sustainable urban development projects possible and to build new urban habitats of the XXI Century. This global understanding of the design is explored in every product from the large scale associated to the territory and the city till the small one related to the design of the architectural space and technical details.

DnA_Design and Architecture

DnA_Design and Architecture 是一家着眼于当代社会、关注各个领域学科、跨越不同尺度的建筑事务所。该公司认为建筑并非是孤立的一个学科，而是触及当代社会的各个层面、各个领域的多维度立体构架。该公司认为文脉、功能以及这二者的相互作用是决定设计、诠释建筑的基本元素，并由此展开研究和讨论，不仅激发每一个项目独特创新的理念，也使设计方案充分适应并融入到当代多样与复杂的社会，最大限度地参与社会变革。

DnA_Design and Architecture 的主要作品有：西溪休闲中心、水岸会所、鄂尔多斯美术馆、宋庄美术馆、吴山专美术馆、大连运动中心、海事大学游泳馆、郑州海联高层、柳州住宅、城市公寓、银川商业区、金华建筑公园 #6 厕所、e 甸园、绥溪档案馆等。

DnA _Design and Architecture

DnA _Design and Architecture is an interdisciplinary practice addressing the contemporary living environment, both physical and social, from scales small to large.

Its approach to projects starts with research and discussion on context, program, and their interaction, which the Company believes are the fundamental elements that will define design and architecture, to adapt, engage, and contribute to society of multiplicity and complexity.

Context, program, and their potential relationship, will cultivate architecture into a multi-dimensional expression and generate new experiment and exploration for users. Architecture will continue to influence and inspire people's contemporary life.

Jorge Hernandez de La Garza

Jorge Hernandez de La Garza 将建筑设计理解为一种具有媒介作用的表达方式，这一方式具体表现在将新科技研究应用到虚拟模型和实体空间中。该事务所热衷于对环境问题、节能问题和可持续设计的研究，通过整合虚拟模型、建筑结构、交通流线、功能以及社会文化价值等各方面的因素，提出一种适应任一案例的空间组织模式。

Jorge Hernandez de La Garza

Jorge Hernandez de La Garza understands architecture as a medium of expression that results in the investigation of new technologies applied to virtual models and physical spaces. A global architectural vision is integrated in every design work from its urban context down to the interior design. The firm is interested in environmental concerns, in saving energy and the implementation of sustainable design alternatives. In the generation of concepts it integrates virtual models, construction systems, circulation, use and sociocultural values in order to achieve a spatial organization that adapts to every new project.

Söhne & Partner

Söhne & Partner 认为建筑必须始终与它所处的环境以及使用这一建筑的人形成对话关系，因此，该事务所总是从项目的周围环境以及特定群体的需求出发，针对不同的场地和客户，以极具组织性和创造性的方法，提出一个个独一无二的专业解决方案。在工作过程中，事务所十分尊重客户的意见和看法，客户可自由地对其工作提出质疑。

Söhne & Partner

Söhne & Partner believes that a building must always engage in a dialogue with its surroundings and the people who use it. The company, based on buildings' surroundings and specific client groups, takes a structured approach to problems, and solve them creatively. The solutions are unique according to different sites and clients. The clients' ideas and opinions are respected in the work process. They are free to question what the company is doing.

MAD 建筑事务所

2004 年，马岩松创立 MAD 建筑事务所，而这一事务所真正进入国际视野始于 2006 年其为加拿大多伦多设计的高层住宅方案在国际竞赛中中标。MAD 一直在寻求一条与环境共存的道路，以东方的自然体验为基础诠释未来主义建筑。MAD 所有的项目，不论是住宅、办公楼还是文化中心，都始终秉承着维护和谐的公共环境和趋于自然的理念，让人们充分体验生活在其中的自在。MAD 现由马岩松、党群和早野洋介领导，2006 年，他们获得纽约建筑联盟青年建筑师奖，并于 2010 年获得英国皇家建筑师协会（RIBA）授予的 RIBA 国际名誉会员称号。

MAD Architects

Founded in 2004 by Ma Yansong, the office first earned worldwide attention in 2006 by winning an international competition to design a residential tower near Toronto. MAD works in forward-looking environments developing futuristic architecture based on a contemporary interpretation of the eastern spirit of nature. All of MAD's projects – from residential complexes or offices to cultural centers – desire to protect a sense of community and orientation toward nature, offering people the freedom to develop their own experience. MAD is led by Ma Yansong, Dang Qun and Yosuke Hayano. They have been awarded the Young Architecture Award from the New York Institute of Architects in 2006 and the 2011 RIBA international fellowship.

Tonkin Liu

Tonkin Liu 是一家一流的建筑事务所，于 2002 年由 Mike Tonkin 和 Anna Liu 共同创立而成。其业务范围包括建筑设计、艺术设计和景观设计，同时，致力于为前瞻性的客户提供与项目场地、使用群体以及当地文化相协调的设计。该事务所认为，每一个项目都是人类与自然关系的具体体现，对自然的深入观察和了解，往往能够激发出一些开阔性的设计和施工技术，因此，在设计过程中，他们通过对自然的关注来展示仿生学的价值以及对自然元素的大胆运用。

Tonkin Liu

Tonkin Liu, established by Mike Tonkin and Anna Liu in 2002, is an award-winning architectural practice whose work encompasses architecture, art and landscape. It provides forward-thinking clients with designs that are finely tuned to the project sites, the people who will occupy it, and the culture that surrounds it at the time. The company believes that each project embodies the relationship between man and nature. Further research and observation of the nature may inspire pioneering construction techniques. During the design process, the company boldly uses biomimicry and the elements of nature in its projects.

TheeAe LTD.

TheeAe 是由建筑折中学派演变而来的缩写词，这一名字本身就代表了创造独特建筑的理念和精神。TheeAe LTD. 自创立以来，就秉持着这一理念，通过新的设计元素，带给人们愉悦的空间体验。除了追求创造性，该团队还致力于满足客户在环境和功能方面的需求，其项目既兼顾了实用性和美观性，还综合考虑了功能、成本以及可持续性。

公司领导人 Woo-Hyuncho 具有十多年在世界各地参与各类项目的经验。他认为，创造性就是一种在大自然的随意性中引入秩序的能力，也正是在这一理念的引导下，他积极寻求有机的设计方法，使建筑随着时间的流逝、功能和用途的变化而产生相应的演变，从而形成一种新的建筑，这样的建筑语言也就是"TheeAe"

TheeAe LTD.

TheeAe is abbreviation of the evolved of architectural eclectic. Its name is ideas and dedication to create unique architecture. Since its establishment, TheeAe has adhered to this concept to bring people joyful spatial experience with new design elements. Besides the pursuit of creativity, the team devotes to satisfy the environmental and functional requirements of clients. All of the projects have well mingled practicality with aesthetics, functional programs, cost and sustainability.

The leader of TheeAe – Woo-Hyuncho has been working with a variety of projects around the world for more than 10 years. As he believes "the creativity is the ability to introduce order into the randomness of nature", his design seeks organic approach which the buildings evolve as they are developed by the change of function and use as time goes by. This eventually will produce the architecture as a new creation. That is the language called TheeAe.